MEDICAL ASTROLOGY: GALACTIC CODE

UNDERSTANDING THE GALACTIC ENERGIES OF THE HUMAN BIOLOGICAL SYSTEMS

TATIANA GALPERINA SCHNEIDER, MD, PH·D

CONTENTS

INTRODUCTION

This book is a result of my long term professional experience as a practicing medical doctor, psychiatrist, as a scientist in the field of biology, as a PhD in immunology and also of my intensive work in astrology.

The famous Greek physician Hippocrates said," A physician without knowledge of astrology has no right to call himself a physician."

Although many physicians and professionals in the field of medicine and biology would hardly accept it, this book is a scientific attempt to prove this statement.

This book is designed to be a guidance to heal using the knowledge of astrology, as well as to make a scientific connection between the astrology and medicine.

It was Greek philosopher Plato who first noted that the planets behave according to mathematically describable patterns. This made it clear that heavenly bodies were not only "higher" in a literal sense (being far from the center of the earth), but also higher on the scale of existence. In contrast to the unpredictable and often chaotic events taking place on earth, the behavior of the planets was governed by rules. Being comprehensible to the intellect, planets and stars were thus considered closer to the Divine and more "real."

Nowadays, medical astrology can be employed with great accuracy as a preventative tool, revealing places in the body where disharmonies may manifest. Medical astrology paints a precise picture of a person's physical, emotional, mental and spiritual conditions. By becoming aware of the body's inherent strengths and weaknesses, one can take the necessary steps to ensure a life of physical and psychological health.

The zodiac is a circle of twelve thirty-degree divisions of celestial longitude centered on the ecliptic. In simpler terms, the Earth's orbit around the sun is the ecliptic. Dividing this huge circle into twelve slices of thirty degrees each gives us the divisions—or "signs"—of the zodiac. Each sign represents the constellation of stars visible in that slice of sky, from Cancer to Sagittarius, which is the closest to the center of the Milky Way galaxy and therefore the most distant from our sun.

Planets carry the energy of particular signs/constellations, and this planetary "rulership" was originally assigned in ancient times but has since been modified to account for the discovery of "new" planets unknown to the ancients. For reasons that will become clear, this book suggests additional rulership changes.

Astrologers have known for some time that the positions of the planets and their associated signs at the moment of birth profoundly affect the biological, psychological, and social makeup of the human being. The cosmic dance of energies is never-ending and ever-changing and leaves its mark on us at the moment we draw our first breath.

What has not been proposed yet is that this constant flux of galactic energies actually contributes to the formation, growth, and health of the organs and operational systems of the human body. The human genetic code, for all its marvelous sophistication and variety, is composed of only four basic components—labeled A, C, G, and T—arranged in different combinations.

Imagine, then, the variety and complexity to be found in the biopsychosocial makeup of a being whose birth, growth, and continuing development are affected, perhaps even determined, by a galactic code composed of twelve constellations, or signs, and eleven planets located in those signs. This occurs in different combinations with each. And like DNA, which, once encoded, remains the same, astrological influences are encoded in our bodies during embryological development and at the moment of birth.

Conventional thought is that these influences are strictly external, but this book will demonstrate that the forces of the signs and the planets exert their influence from within as well as without. While it has been known for some time that certain constellations rule specific body parts, this book goes further, showing that each sign is represented in the body by one human system and that each human system organizes itself into the division of three to form a bigger system or element of the body.

The body also contains an imprint of the solar system, which we call in the book "the inner human solar system." The inner solar system regulates all the developmental stages of life. Each of the planet energy centers finds a location in the human body where it can express itself most strongly. Together the inner planets make an imprint of a human head and body. For example, the energy of Mercury is felt in the frontal part of the head.

As above, so below; each body represents the zodiac in the sky and contains the solar system that represents a small homunculus interacting with a big person of the Zodiac signs. The solar system directs the evolutionary development of life on Earth.

At the time of birth, the planets make an imprint on the zodiac in the body. The distribution of the planets at the moment of birth, each ruling a particular sign and carrying its energy, creates a blueprint of energies in the human zodiac that is different for everyone, imbuing each individual with a unique energy field. This one-of-a-kind mix of energetic interactions is the "Galactic Code." The Galactic Code exerts a combination of influences that is also unique for each individual.

As in poker, the cards one has been dealt strongly influence the way future events unfold—but those who know the game and its rules enjoy a distinct advantage over those who have little knowledge of the cards or what they represent. This book is an effort to provide the advantage of astrological knowledge.

Let us turn now to the matter of how the signs and the planets produce these human energies and how those energies influence our physical, intellectual, emotional, and social development.

FOUR ELEMENTS AND THE DESIGN OF THE HUMAN ZODIAC

The twelve signs of the zodiac are like factories, creating different energies. Those energies organize an internal zodiac that arises in the body. I am presuming here that the energies forming the internal zodiac reside in the twelve systems of the human organism, each system manifesting the energy of one sign.

The idea that the energies of the zodiac are contained in different parts of the body was first codified in the ancient system commonly called the "Zodiac Man." Sketches of the Zodiac Man appeared frequently in formal expositions from the Middle Ages, showing where the astrological signs were believed to have influence over the body and its health. As per this Zodiac Man image, the zodiacal constellations appear to *code* specific parts of the human body.

While doing the readings, I arrived at the conclusion that the signs of zodiac also take part in the biological design of the human body. My theory assigns the psychological meaning of the signs to the energies of the main biological systems of the body. Those systems provide the biological substrate for the psychological energies we use in our daily life.

I presume here that human systems are acting as the reservoirs of cosmic energies. The unique interplay of those energies is brought to the energetic field of each person by the planets of the solar system as per the natal chart. That makes a blueprint, activating certain places of the zodiac signs in the body. At the same time, each planet brings the energy of the zodiac sign over which it rules.

The twelve signs of the zodiac belong to the four elements: earth, water, fire, and air. I assumed the human systems that carry the energy of the same element belong to the same three-part system of the body. In this way, the twelve signs of the zodiac are represented in the human body by the four elements of the zodiac, which correspond to the four human systems. Each of those systems includes three parts corresponding to three signs of the same element.

Which systems do the zodiac elements represent and what three parts do they include? The three parts of a human system are: the part with a task, which is the leading part; the part that is in the process of fulfilling this task; and the ending part, which carries a resolution and communicates the results to the leading part.

In the human body, the leading part gives the command that goes to an organ that then undertakes the process of fulfilling the command. The feedback from this working organ goes back to the part of the body that analyzes and processes the results. It is important for the leading part to know the results in order to course-correct or end the accomplished task; this is called biofeedback. The ending part of the body continues the biofeedback and communicates with the intermediate part. The intermediate system communicates with the leading part to hear the changed task again.

Among the signs, the leading part is the cardinal sign, the intermediate part fulfilling the process is the fixed sign, and the part analyzing and processing the work is the mutable sign.

The mutable sign carries the end result of the system's work and is transformed, in contrast with the cardinal and fixed signs, as it carries a transformed consciousness to the next system representing the element with a different function to be fulfilled. The functions of those parts of the body are assumed based on the system of the Zodiac Man, on the location of the body parts themselves, and on the functional activity, chemistry, and structural organization of the parts. Those findings will be further analyzed in subsequent chapters

THE FIRE ELEMENT AND ITS PARTICIPATION IN THE DESIGN OF THE HUMAN ZODIAC

The fire element is represented by three signs: Aries, the cardinal leader sign; Leo, the intermediate fixed sign; and Sagittarius, the ending mutable sign (Picture 3). The fire element of the zodiac builds the fire system of the human body, which represents the emotional and expressive human consciousness. The correspondence of the zodiac signs to the parts of the human systems is the main presumption and foundation of this book.

Aries represents the limbic part of the human brain, which regulates a person's emotional state. As a cardinal sign, Aries is responsible for storing the element that controls the fire system: oxygen. Oxygen is held in the nose and the nasal cavities. As oxygen is found only in the form of particles and not in waves, those born in the air signs tend to be private people.

The reassigned ruling planet for the sign of Aries is the moon. The moon is the planet that experiences and transmits the emotions. Unlike other planets, the moon doesn't act or react; rather, it accumulates emotional experiences. The moon radiates light, which is a feature of a fire system. It is also responsible for the female hormones, which influence the emotions.

Aries is represented as a ram—an animal well-known to experience emotions—with its horns pointing to the third eye.

Leo is the second sign of the fire element. It represents the unconscious parts of the brain: the parietal, temporal and occipital lobes. These unconscious parts of the brain belong to the intermediate part of the fire system ruled by the sun—and allow the individual to experience the creativity and potential of the true self. The sun possesses an enormous unconscious potential, which makes it difficult to fully express oneself and feel successful in life.

The lion (representing the Leo sign in the picture) is known through many stories as the ruler of the animal kingdom.

The next mutable sign, Sagittarius, represents the hip, large muscles, the motor nuclei of the spinal brain, and the metabolism necessary for the whole body to move and function, including the energy-producing organs such as the liver. It provides the capacity to move and express oneself. Sagittarius is the ending part of the task of emotional expression. Jupiter, being its ruling planet, has a huge potential to expand, travel, act as a teacher, and play different roles.

It is interesting that the three planets ruling the signs belonging to the fire element are involved in rhythmic motions around the sun, during which they spend a consistent amount of time in each zodiac sign.

The fire system seems to be the system that explores the emotional energies of the signs. Sagittarius is represented in the picture by a man with an arrow, aiming to express himself in the world. Aries, represented by the moon and the Earth, reflects the emotional consciousness, the ability to care for oneself and for Earth. Leo, represented by the sun, reflects the emotional consciousness of the brain—its ability to respond to emotions.

THE WATER ELEMENT AND ITS PARTICIPATION IN THE DESIGN OF THE HUMAN ZODIAC

The water element is represented by three signs: Cancer, the cardinal leader sign; Scorpio, the intermediate fixed sign; and Pisces, the ending mutable sign (Picture 2).

Cancer represents the super-conscious part of the brain, located between the unconscious parietal lobe and the conscious frontal lobe. It may regulate the main pineal gland, which is responsible for awareness of night and day. The role of Cancer is organization, recognition, and acceptance. The stomach plays this role by absorbing foreign material.

As a cardinal sign, Cancer is responsible for holding and storing the element regulating the water systems, which is water itself. Water can be in a simple form that represents a particle as well as in a crystallized form that has the ability to form protonic H+ waves. People born in the water signs have a collective nature but also take a privately creative approach to life. Cancer people are cordial and generous and have strong organizational skills.

The planet that rules Cancer may be a star outside our galaxy, possibly Sirius, which points to the middle of the Cancer constellation. A fixed star, it doesn't move and always points at this direction.

Cancer would move in an opposite direction of the Galaxy that may point to the location of the ruling star.

As mentioned, Cancer people resolve a conflict between collective and individual energy by being an organizing force for friends and family while at the same time maintaining a private space for themselves. The collective energy comes more easily for them; carving out a private creative place for themselves is more of a struggle. Cancers wear a protective shell to maintain that private space.

Scorpio represents the reproductive system. In the biological sense, it's a collection of different types of antigens produced by the reproductory organs. The antigens are used for genetic selection. In life, Scorpio is strong on emotional recognition, sensitivity, and intuitive skills.

The ruling planet is Pluto. The planet of transformation, it shifts energy into new directions. Pluto represents a transformational revolutionary stage marked by the appearance of new antigens necessary for the organism.

As with all the water signs, Scorpio is represented by an animal covered with a shell. Scorpio is an insect species where the female eats the male, incorporating his antigens into her system. It represents the desire to combine collective life with a private life.

Pisces represents the lymphatic system, in which the processes of lymphocyte proliferation and education occur. It regulates the process of the lymphocyte maturation in the thymus.

Psychologically, Pisces tries to accept the multi-faceted qualities of others as well as support the individuality of oneself.

The ruler of Pisces is Neptune, surrounding whom is a cloud of trans-Neptunian objects.

Pisces is a sign represented by two fish swimming in different directions, symbolizing that they swim close to one another yet have different destinations and belong to different systems. One is being preserved, another eliminated. The fish are covered with shells, which show their struggle to combine the collective intentions with a private life.

The systems of water signs are regulated by the element of water. Cancer regulates the pineal gland and is connected with the regulating of daytime light. As a cardinal sign, Cancer holds and stores the controlling element water in the stomach. The planets that rule the water systems—Neptune and Pluto—are located in the center of the galaxy, while the star Sirius remains outside the galaxy.

The water system aims to multiply the genome as well as solidify the processes of nutrition and protection in order to accept the variety of life and combine the personal with the collective. As water clears out all the differences along its path, the systems that belong to the water element works toward multiplying, recognizing, and improving its own material and subsequently the genotype.

THE AIR ELEMENT AND ITS PARTICIPATION IN THE DESIGN OF THE HUMAN ZODIAC

The air element is led by Libra, the cardinal sign; Aquarius, the fixed intermediate sign; and Gemini, the ending mutable sign (Picture 4).

Libra starts the task of relationships. Within the organism, this is the task of the vegetative nervous system and the twelve nerves that come from the base of the brain. As a cardinal sign Libra holds and stores the element of the system—the cations that initiate the electromagnetic impulses in the body (also called "the wind"). The cations are stored in the spinal fluid. As the electromagnetic impulse has a collective quality, people born in the air signs have a collective life approach.

Libra represents the relationship of the nervous system with different organs and the ability to distribute adequately to oneself and others. The ability to react differently to each organ by using the law of biofeedback comes into play in a major way. Libra is a relationship sign—it wants to make connections and give to everyone. At times, it struggles with how to distribute its attention and sometimes forgets or diminishes itself.

Mars is the reassigned ruling planet for the sign of Libra. Mars rules this sign because of Libra's connection with action. Mars is the personal planet that can provide the action. Mars in Libra is known to respond to everyone by using the rule of biofeedback, respectfully balancing its actions with the actions of others.

Aquarius represents the vast body of the vessels with neuronal endings that belong to the neuro-vegetative system. People whose sun sign is in Aquarius harbor a vast amount of knowledge. They think for the whole body and are also eager to experience the act of giving to others.

The ruling planet is Uranus, which is connected with knowledge and social activity. Aquarius transfers its knowledge by means of a general and global approach.

Aquarius is shown pouring water because this sign is located closest to the water element Pisces, which is ruled by the planet Neptune. Aquarius fills the water systems of the blood vessels with lymphocytes of Pisces and their memories and supplies the whole body with neuro-regulation. Uranus is in close chemical and structural connection with the planet Neptune.

Gemini represents the executive brain, with its frontal lobes, as well as the chest and the arms, including shoulders and hands. This central nervous system is conscious and able to quickly direct activities, especially the parts of the chest that are represented by this part of the brain. The executive brain analyzes how well the work of communication and cooperation has been accomplished. Gemini thinks quickly. The ruling planet is Mercury, which directs thoughts and communications and oversees a conscious mind.

The planets of the air element are all relatively small.

Gemini carries the information of the mind, Libra is on the border of the central nervous system—it carries the information of the mind and the body, Uranus – of the body – with its blood vessels filled with neuro-receptors. In other words, the air system represents the neuro-vegetative system, with centers in the brain that direct neuronal activity in the peripheral parts of the nervous system, the nerve endings that accomplish each function, and the executive part of the nervous system that analyzes and corrects those functions.

Swiftly carrying ideas and words in the same way that air flows, the neuro-vegetative system has an air-like quality as it carries electrical charges from nerve to nerve. This neural network is the system that accomplishes all the body's internal functions.

THE EARTH ELEMENT AND ITS PARTICIPATION IN THE DESIGN OF THE HUMAN ZODIAC

The earth element is represented by three signs: Capricorn, the cardinal sign; Taurus, the fixed sign; and Virgo, the ending mutable sign (Picture 2).

Capricorn represents the bones and the bone marrow. The bones are the foundation of the body. As a cardinal sign, Capricorn holds and stores the element that controls the earth systems: light. Light is stored in the canals of the bones. As light may be in two forms—waves and photons—people born in the earth signs take both private and collective approaches to life. Usually, the private personality is more natural to them, and they struggle to be social and to combine both qualities.

Capricorn works on designing the proper project and takes full responsibility for and control of designs in the material world.

The ruling planet is Saturn, the big and structured planet responsible for self-consciousness.

The goat pictured in the sign of Capricorn is known for its love of jumping and running, which is connected to the work of the bones. The goat is also known for its desire to do things in its own way. Capricorn likes to rely on himself and his bones and validate himself.

Taurus, the fixed sign of the earth element, regulates the sensory organs, which supply the person with sensory input and visual images. The main nervous cells that Taurus operates with are the cones and rods in the eye that react to light.

Taurus is known for its desire to do his best in life, to distinguish right from wrong, and to do things in a proper manner. The ruling planet is Venus.

In the picture, Taurus has a bull's face that represents the face and the senses. In Greek mythology, the goddess Venus appears once as a bull.

The mutable sign of the earth element Virgo regulates the cerebellum. The cerebellum consists of chains of nervous cells that receive multiple stimuli. There are supporting cells and fibers that help to analyze and interpret the right stimulus and make the right choice. It works to extend, receive, and monitor many signals. The most highly organized cells of the cerebellum are the Purkinje cells—the highly organized nervous cells of the brain cortex of the cerebellum—which choose one signal from among the many arriving through different fibers.

Virgo works on developing the ability to analyze how well things are done in order to make right decisions.

The ruling planet for Virgo is Chiron.

Virgo represents the virgin from Greek mythology. The virgin Persephone was the daughter of Zeus (Uranus) and wife of Hades (Pluto). She was abducted by Hades and spent six months on Earth for every six in the underworld. The image shows that Virgo represents the human being who is approaching divine wisdom.

The earth element is about growth, ability to stand for oneself, validate, behave, and make proper decisions.

So in the system of the human zodiac, the four elements regulate four different systems of the human body: the emotional (creative) motor-metabolic system (the fire element); the neuro-vegetative system (the air element); the antigen-recognition systems, combining the digestive, reproductive, and protective systems (the water element); and the designing systems—the skeletal and cerebellum systems and the eyes (the earth element).

The signs that code the human body interact with each other through the glands, the nervous system and their elements and are in constant cooperation. Each of the four systems is regulated by its own element. The element is accumulated and stored by the cardinal sign, used by fixed sign, and used and disseminated by the mutable sign.

Each element is based on the foundation built by another element. The sequence of the elements is moving clockwise (more in Chapter 10). The significance and the interplay of the elements will be discussed at greater length in further chapters.

ZODIAC SIGNS AND THE QUALITIES OF THEIR ELEMENTS

As described above, my proposition is that certain systems in the body represent the elements of the zodiac ecliptic, and in accordance with this, the body carries the energies of the zodiac signs in its systems.

Cancer is considered to regulate the top of the brain and so is at the top of the chart. As the star Sirius points to 50 degrees of the Cancer sign, this placement at the top of the chart makes the chart symmetrical.

By the same token, Capricorn—known to regulate the legs and the knees—came to be positioned at the bottom of the chart (Picture 1).

As the chart was thought to represent the person, the system in the human organism that represents certain elements was thought to make a triangle in the body, similar to that made by the signs of the zodiac on the ecliptic.

In the previous chapter, I described the parts of the systems that were found, based on their location and function in the body. Three signs belonging to one element make up one system, which reflects the location and function of this element in the human body.

The signs of each element make a particular triangle on the human ecliptic that corresponds both to the celestial triangle and to a similar one composed of systems in the human body. In other words, it's possible to imagine the human ecliptic as carrying the energies of the twelve human systems, each one occupying 30 percent of the area on the human ecliptic (Picture 1).

PICTURE I· THE HUMAN ECLIPTIC

Comparing the human ecliptic to the human zodiac, we can imagine the human systems carrying the energies and forming certain shapes, which is similar to what happens in the celestial constellations. It could be said that the systems of the body reflect the constellations of the celestial zodiac.

The zodiac signs on the human ecliptic are the schematic representations of the energies of human systems, just as the zodiac signs on the celestial ecliptic represent the energies of the constellations.

THE WATER ELEMENTS

The cardinal sign of Cancer is at the top of the head; that's why the triangle of the water signs starts at the person's head and descends toward the body (Picture 2). In the body, it connects to the signs located at the same distance from the sign of Cancer, so it's a downward-pointing triangle. It's interesting that because Pisces regulates also heels and some metatarsal bones, the water elements form a prism instead of a triangle, embracing the person's whole figure. The triangle makes the upper half of the prism.

The triangle of the biological systems that are regulated by the water signs reflects the triangle of these signs on the zodiac ecliptic (Picture 2A).

PICTURE 2: THE SIGNS OF THE WATER TRIANGLE ON THE HUMAN ECLIPTIC

PICTURE 2A: THE SIGNS OF THE WATER TRIANGLE ON THE HUMAN BODY

THE FIRE TRIANGLE

This triangle is shorter than the others, ending at the person's hip (Picture 3). The triangle goes sideways from the sign of Aries, which is located on the left side, toward the right side.

The signs of Leo and Sagittarius are also symmetrical toward the vertical axis. The sign of Leo is located next to the sign of Cancer.

The triangle of the human systems that are regulated by the fire signs is similar to the triangle of these signs on the human ecliptic (Picture 3A).

PICTURE 3· THE SIGNS OF THE FIRE TRIANGLE ON THE HUMAN ECLIPTIC

PICTURE 3A· THE SIGNS OF THE FIRE TRIANGLE ON THE HUMAN BODY

THE AIR TRIANGLE

This triangle goes sideways from the cardinal sign of Libra, starting at the right side and going left (Picture 4).

Libra regulates the medulla oblongata. Aquarius and Gemini correspondingly regulate the ankle and the nervous receptors and the frontal lobes. Two signs on the left side are symmetrical toward the vertical axis. The sign that comes closest to Cancer is Gemini.

The triangle of the biological systems that are regulated by the air signs reflect the triangle of these signs on the zodiac ecliptic (Picture 4A).

PICTURE 4: THE SIGNS OF THE FIRE TRIANGLE ON THE HUMAN ECLIPTIC

PICTURE 4A: THE SIGNS OF THE AIR TRIANGLE ON THE HUMAN BODY

The triangles of air and fire move sideways toward each other and regulate the interactions of emotions and relationships.

Those are two helpers in the interaction of water and earth: God and the human being.

We can imagine that in ancient times, the fire and the air elements were described as the two trees in the Garden of Eden. The fire tree, the one of good and evil, attracted more attention. The air-element tree was the one bearing the name the Tree of Life.

The triangle of earth is directed straight upward to meet the triangle of water. As Pisces codes also the heels and toes, the element of water covers the whole person.

The triangle of air, in its development, travels to the more conscious left side of the body, while the triangle of fire travels to the more unconscious right side. (In the organism, the left side represents the front of the body, the right side the back.)

The air element's travel reveals the development of human

communications; the development of the element of fire reveals human emotional individuality.

THE EARTH TRIANGLE

Capricorn is the cardinal sign of the earth element and regulates the knee, which is where the earth triangle begins (Picture 5).

At its top, the triangle comes to the person's head and the signs located an equal distance from the sign of Cancer (and regulating the eyes and the cerebellum). So the triangle is directed upward.

The triangle of the human biological systems that are regulated by the earth signs is similar to the triangle of these signs on the human ecliptic (Picture 5A).

PICTURE 5 THE SIGNS OF THE EARTH TRIANGLE ON THE HUMAN ECLIPTIC

PICTURE 5A· THE SIGNS OF THE EARTH TRIANGLE ON THE HUMAN BODY

As seen on the parallel images depicting the elemental triangles, the shapes of the triangles connecting the signs of a certain element correspond to the triangles connecting the body parts they regulate.

The correspondence of the triangles connecting the signs and the systems of the human body regulated by these signs reflects the existence of the human zodiac carried by human tissues and running in the same sequence as with the ecliptic in the sky. The organs carrying the energy of the zodiac signs also follow each other in the same sequence as the signs on the celestial zodiac.

THE AXES ON THE HUMAN ZODIAC AND THE INTERACTIONS OF SYSTEMS LOCATED ON THE OPPOSITE SIDES OF THOSE AXES IN THE HUMAN BODY

There are six axes connecting the signs of the zodiac on the human ecliptic (Picture 6).

PICTURE 6· THE AXES OF THE HUMAN ECLIPTIC

The earth signs connect with the water signs, forming harmonious relationships, but they form non-harmonious relationships with the air and fire signs.

1· CARDINAL VERTICAL AXIS

The main vertical axis of cardinal signs Cancer – Capricorn is about the relationship between human and God. The activity of the human on the earthly plane opposes the organizational family activity of God.

The water and earth signs oppose each other but never come into direct contact. They can achieve contact only through the two helpers: the elements of air and fire. This might represent two trees that were planted in the Garden of Eden. Only one tree was useful for the person and was said to bring eternal life. The other was forbidden.

In the body, the element of earth represents the bones and, later, the nervous system that connects different parts of the organism. The ultimate goal of proper development of the nervous system is to raise the person to the same level of development as that of the divine mind.

As the fire element represents the emotional system, so the air element represents the vegetative nervous system. As the water elements have both an individual and collective consciousness, the earth element has analogous and digital control achieved by light. The analogous consciousness of the earth is mostly programmed in a collective way, the digital in an individual way.

As the person has to walk a path to come to God, he needs to walk through half of the zodiac. As we will see, the path is directed clockwise and leads through the development of relationships and of the central nervous system. The tree of the air element and of relationships is the same one that may promise eternal life.

2· CARDINAL HORIZONTAL AXIS

The cardinal horizontal axis represents the interaction between the signs of the air and fire elements, which are the intermediates between human and God. As the vertical cortical axis is the axis of the months of solstice, the horizontal cardinal axis is of the months of equinox: Libra and Aries.

Libra regulates the medulla oblongata in the base of the brain, which is the main path of emotional interaction and reaction-correction. As Aries regulates the emotional consciousness of the person and develops in the early stages of life, it regulates and balances relationships and is the main factor in balancing and stabilizing the emotions. In human development, the parents and close relatives make themselves the main factor in the developing emotional brain of the child.

Two major cardinal signs program the structure of emotions and relationships and meet each other for interaction and stabilization. Interaction is so important that it is happening on the major neurologic paths in the base of the head. This is the place of the connection of the equinoxes. The signs on the horizontal axis delineate the condition of consciousness on Earth. Aries is the state of individuality, Libra the state of interaction with another being.

The signs above the horizontal axis regulate different brain structures located higher than those regulated by Aries and Libra. The five signs located above the horizontal axes oppose the five signs below the axes, signifying interactions with the divine as opposed to those with humans.

THE UNIQUE DISTRIBUTION OF THE SIGNS AND THEIR RULING PLANETS IN THE HUMAN ZODIAC

My concept of the connection between astrology and the body comes from biological science and the structural and functional anatomy of the person. In the previous chapter, I described the broader system of how the signs of the zodiac "code" the body with their individual functions. The codes described were based on the correlations of the elements of the zodiac with particular systems of the human body and with the signs of the zodiac.

After I found correspondence with bodily systems, the signs of the zodiac organized themselves into the internal zodiac. When the ecliptic of the celestial zodiac meets the internal zodiac ecliptic around a human being, the signs go in the same complete sequence, corresponding with parts of the body.

On a circular diagram where the signs line up along the ecliptic around the body, they align with systems and body parts while the person is standing on his feet with his head up (Diagram 1).

I have put the summer months on the top of the chart, corresponding with the head, and the winter months at the bottom, corresponding with the feet.

The elongation of the human body relates to the elongation of the ecliptic in the sky; the summer and winter signs are located on the more distant parts of the zodiac ecliptic. The months of the equinoxes are located on the side parts of the ecliptic, along the main horizontal axis of the astrological chart. On the body, that corresponds to the level of the neck and the path along the base of the brain.

In this way, the ecliptic in the sky mimics the ecliptic around the human body and may in fact be coding its energies into human beings. The existence of the human ecliptic is provided and proved by specific systems in the body.

The different transient movement of the planets on the celestial ecliptic accomplishes the esoteric work of changing, transforming, and adjusting the energies on the human ecliptic. The positions of the planets at the moment of birth and their interaction contribute to the state of the psychological and biological health of the person.

The planets of the solar system transmit the energies of the signs of the zodiac to the person's body. An important concept of this book is that this happens in such a way that each planet carries and transmits the energy of one particular sign that it rules and doesn't carry or transmit energies of different signs. This is a particular quality of the Galactic Code.

Because in some current classifications Venus rules two signs, some changes to the rulership system are offered:

I. CLASSIFICATION OF THE SIGNS ALONG THE HUMAN ZODIAC

Of the twelve signs:

- Three signs are arranged on the very upper level, on top of the human ecliptic, and regulate the brain. These are therefore "brain signs."

- Four signs group around the person's face and the back of the head, with two signs on each side. These form the "human individuality" signs. They code the consciousness and individual qualities of the person: the perception and experience of the senses (Taurus) and emotions (Aries) and his way of responding, balancing (Libra), and analyzing them (Virgo).

- Two signs are located on each side of the body, back and front. These are the "body signs," formed by signs belonging to the water element. They carry the reproductive and protective function, Scorpio from the back side of the person, Pisces from the front.

- Three signs are grouped along the joints of the leg—the "leg signs." They are located on the hip, on the knee, and on the ankle. These code the general functions of the body: the peripheral nervous system; bones and bone marrow; and muscular vascular- metabolic systems. These are coded by the signs of Aquarius on the ankle, Capricorn on the knee, and Sagittarius on the hip.

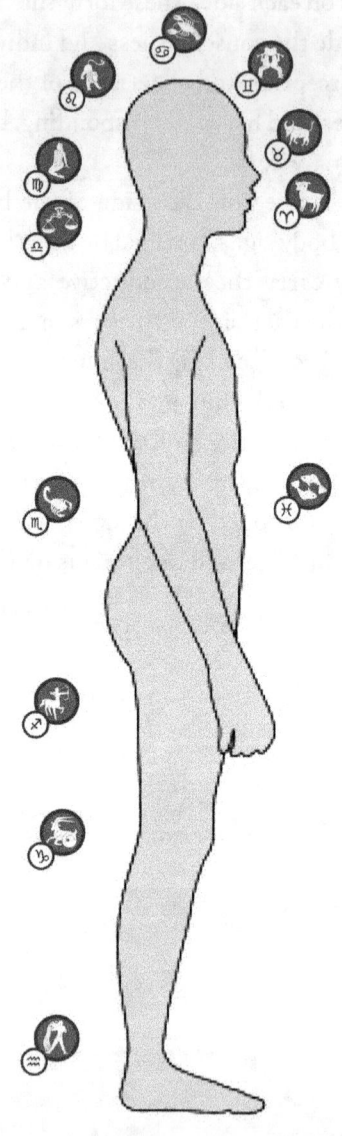

DIAGRAM I. THE DISTRIBUTION OF THE ZODIAC
SIGNS ON THE HUMAN ECLIPTIC

THE THREE 'BRAIN SIGNS'

The three signs that are grouped as a diamond at the top of a person's head *regulate* parts of the brain and chest/organs. The signs that regulate the different parts of the brain are, from front to back, Gemini, Cancer, and Leo.

Gemini codes the conscious parts of the brain: the frontal lobes that process information, mostly in the manner of conscious thoughts. Gemini also creates the chest, arms and hands. The representations of all the parts are located on the frontal lobes, so any impulse that comes to the person's frontal lobes can be immediately used by the hands, making them a useful instrument in many professions and in life. Gemini is consciousness expressed in thought and communication. Gemini is located on the left side of Cancer, which organizes the work of designing the body for the whole zodiac ecliptic. Gemini is ruled by the personal planet Mercury.

Leo codes the unconscious part of the brain—the parietal, temporal and occipital lobes—and is ruled by the sun. It stands on the right, western side of Cancer.

The sign of Cancer is the lead part of the water element and the "brain signs." It's located in the middle of the brain and on top of the ecliptic. In itself, it may regulate all parts of the conscious, unconscious, and super-conscious mind. The super-conscious mind includes the pineal gland, which regulates the major hormonal systems, the day-night switch, and the body's circadian rhythm. Cancer also regulates the stomach, which is known to react to emotions and thoughts and can be part of the unconscious mind. The stomach also belongs to the systems of the water element, with its job of analyzing and accepting foreign bodies in the form of nutrients.

Cancer is in the sky in the middle of the summer, when the sun is at the midheaven, and coincides with the summer solstice. The constellation points toward the brightest star in the sky: Sirius, which is located at fifteen degrees of Cancer.

Cancer organizes the functions of the signs on the ecliptic, as it sits on top of the chart and is known to have the qualities of uniting and organizing. The super-conscious part of the brain that it regulates—the superego—is connected with the perception of truth, and because of this, Cancer is able to direct other signs on the ecliptic. The planet ruling the sign of Cancer has to have the organizing ability to manage the control and organization of the body.

The moon relates to emotions and regulates the hormones, and it is connected with the Earth. That is why the rulership of the moon in this book is changed for the sign of Aries—the sign that is specific to the emotional mind pertaining to the Earth.

The planet ruling the sign of Cancer remains unknown. It may be a fixed star from another galaxy, pointing to Cancer. As mentioned, it could be Sirius.

THE FOUR 'HUMAN INDIVIDUALITY' SIGNS

The four "human individuality" signs—Taurus, Aries, Virgo, and Libra—regulate different parts of the brain.

Taurus codes the eyes, the throat and the senses of the face. In its connective-tissues regulation, it is especially associated with the cartilage of the throat and with the vocal cords, which are sensitive to hormonal regulation and designed to create sound. So Taurus codes perception and recognition through the senses as well as the elements of the throat.

Taurus creates the eyes with their multilayered nervous cells, a structure tuned to react to light and to choose one signal from among many. The eyes represent the conscious approach of the human in creating his own world.

The role of the earth element is to design and consciously choose from among different stimuli to create its own way in perception of image. Venus is the ruler of the constellation of Taurus.

The senses are the major instruments that help in the creation of the emotions. Emotions and relationships are the two major distinguishing features of human development.

Aries regulates the limbic system—the emotional brain, which rises based on the foundation of the senses and visual images. It is connected with the nose and the nasal cavities containing oxygen. Oxygen, with its hemoglobin, could be the major carrier of emotions and emotional memories.

The reassigned ruling planet for the sign of Aries is the moon. The moon is the planet that experiences and transmits the emotions. Unlike the other planets, the moon does not act or react; rather, it accumulates and feels emotional experiences.

The moon allows the person to feel the emotional energies of all signs on a monthly basis. The moon in Aries in the natal chart allows the person to maximally perceive emotions. In other words, it makes the person sensitive to emotions. All the fire signs are about feelings and self-expression, and like the luminary sun rules Leo, the luminary moon rules Aries.

It seems that the feminine planets—Venus and the moon—rule the front of the person's head, while the back has the male rulership of Mars. The base of the brain becomes the place of interaction between the feminine and masculine influences—feelings and emotions, response and action.

When adding the influence of the planet Mercury, the interaction between the personal planets is that of emotions, thoughts, feelings, words, and deeds.

The base of the brain, where the two opposing equinox sides meet, becomes the place of interaction and the contact point for the major personal planets and the male and female. The interaction occurs along the main horizontal axis of the human chart Aries – Libra (Diagram 2).

The sign of Libra, which regulates the medulla oblongata, descends through the twelve nerves to the chest, stomach and pelvis and achieves the nervous regulation of all of the organs. It's responsible for the relationships between the body and its organs. Libra represents the parts of the brain that perform the regulatory functions of the vegetative nervous system, which is responsible for relationships with the organs. This regulation is done partly through the control of the nervous receptors of the vessels and the smooth muscles of the body (Aquarius).

Because of Libra's relationship to the action and interaction among the parts of the body, Mars—the major planet of action—is proposed to have rulership over the sign of Libra (Diagram 2).

As neuronal regulation is based on biofeedback, the more the organ responds to stimulation, the more there is nervous reaction to its regulation. Biofeedback is the prevalent control method of the body/organ work. Mars behaves in a similar way psychologically. The nature of the response regulates the actions of Mars. The bigger is the response, the greater is the activity seen from Mars.

From the psychological point of view, Mars in Libra was always seen to have an almost passive-aggressive nature, especially in difficult household situations. In relations with friends and colleagues, Mars in Libra has an active and thoughtful nature, allowing people to maintain thoughtful relationships and good careers at work.

Aries and Libra are on the same axis in the astrological chart. Relationships in a parenting household, and parents' reactions, form the emotional spectrum and experience of a child. Once the emotions of the person are formed, on the other hand, they determine the person's future relationships and spectrum of interactions in the world.

As the moon regulates the response of the human emotional brain, Mars shows actions and responses that are also emotionally based. The better the response to action, the more action is taken to induce further response. An intense reaction sometimes induces a response that cuts the reaction. These are neurologically carried reactions of the body.

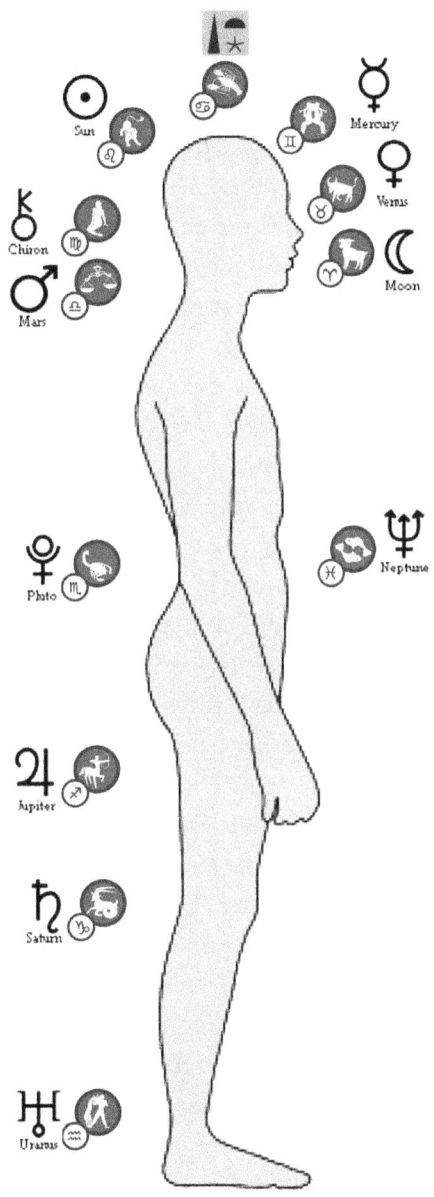

DIAGRAM 2: THE DISTRIBUTION OF THE SIGNS AND
THEIR RULING PLANETS ON THE HUMAN ECLIPTIC

The signs on the Aries-Libra axis and above it regulate the consciousness of different parts of the human brain.

Virgo and Taurus are the earth signs responsible for the brain tissue based on transforming light. Taurus regulates tissues of a very high and delicate nature: the multilayer structures of the nervous cells of the eyes. Virgo codes the nervous tissue of the most ancient part of the brain, the cerebellum, which is known for its extremely complicated structure. Other than being responsible for balance and motor stability, its function remains largely unknown. Its structure is far more complicated than its known functions would seem to warrant.

The Purkinje cells are probably capable of transforming the light, as the rods and cones photoreceptors of the eyes can do.

The Purkinje cells choose one from signal among the very many that arrive to them through different fibers. In this way, the cerebellum undertakes the complicated analytical process of choosing the proper signal from among the many coming in at any given time.

The cerebellum is a highly differentiated part of the brain. The ratio of the cerebellar cortex to the neocortex is a constant 3:6 among all species that have them. (Most don't.) The whole segment of the cerebellum is of a separate origin from the other parts of the brain. The multilayered structures of the cerebellum resemble the multilayered nervous structures of the eye. In both systems, there is a differentiation of signals and the possibility of choosing one signal from among many. The signal starts to represent the person's current reality. This way, through the mechanisms of the earth-element signs Taurus and Virgo, the person has the tools for creating and choosing his or her own reality.

The mechanisms that are presented by signs belonging to the earth element remind us of the stories presented by the major spiritual books. The Mahabharata shows a struggle between a thousand *kurus* (the thousands of bad habits of human nature) and five *pandavas* (the five human senses) and tells how the *pandavas* and *krishna* (the sixth sense, that comes from the brain) may overcome and conquer *kurus*. The structures of the earth element teach the person to perceive and choose in a highly organized, graceful, and disciplined way in order to create a more joyful reality. A possible function of the Purkinje cells is to create a human consciousness that is of a light nature.

Virgo is located just one sign away from the main sign of Cancer, and its purpose is to overcome the unconscious. The same task is presented to the sign of Leo (ruled by the sun, representing the unconscious creative parts of the brain). Taurus is also located only one sign from Cancer, on the left side, which has more conscious signs than the right side. The senses of the person consist of more conscious nervous structures. Gemini, as well, represents the conscious part of the human brain—the frontal lobes.

By becoming more conscious, the sign of Virgo (representing perception and well-being) helps humans to use more intellect and therefore enables them to follow the more conscious path that leads to God or ultimate truth. Chiron, a recently discovered planet, has been assigned as the ruler of Virgo. Chiron has a possibility to move, which may be interesting given that the person's consciousness may be locating in different dimensions or worlds.

Virgo plays a dual role. It fulfills the part of a person designing his own world and making healthy choices, and it also becomes part of conscious communication and in this way stops being separate and becomes one with God. As a deva—a virgin—Virgo belongs both to the earthly spheres and to the celestial spheres, as it is ruled by Chiron, a centaur, and may be traveling between both realms.

Chiron's cycle around the sun takes fifty-one years and represents the maturity of the brain. Through Virgo, a person can accomplish the task of self-growth. The human brain develops toward being able to perceive and encounter a better, healthier, and more organized life. Growth in this direction leads to bliss, recognition of joy, and service in life.

THE TWO 'BODY SIGNS' BELONGING TO THE WATER ELEMENT ON THE SIDES OF THE BODY

The two water constellations at the back and front of the body rule the reproductive system (Scorpio) and the protective system (Pisces). The element of water has been designed to regulate the most primary functions of living systems: digestion, reproduction, and protection. It has regulated the most ancient forms of living beings.

All the water signs represent the systems of recognition of the self from other. These systems have been present since ancient times as the most basic and necessary for all living beings and have been constantly transformed into more complex and organized structures. The sign of Scorpio represents the reproductive system and the merging of the self with others in the forming of intimate relationships. The sign of Pisces represents the lymphatic system, which organizes the work of multiple forms of blood cells.

Jesus who came in the millennium of Pisces taught his disciples to be compassionate, teach, and live in a cooperative and loving way. He said that this way of living would protect them from all evil. Thus this represented the immunity function of the lymphatic system as directed by Pisces. Good immune function enables US to "walk" the protected way along the ecliptic toward the blessings of health and prosperity.

It is interesting that Pisces may also code the heel and the toes. The age of Pisces, in which Jesus washed the feet of his disciples, may have started the human path toward the divine.

Neptune and the trans-Neptunian objects, including Pluto, rule the water signs, as those in turn represent the major and basic systems for all living beings. Pluto rules Scorpio and the process of reproduction and transformation of the genotype. Neptune rules Pisces and the process of healing. Cancer, as mentioned earlier, is ruled by a separate star, possibly Sirius.

THE THREE 'LEG SIGNS'

The emergence of the new individual begins with the creation of the leg of his independence. It is occupied by the signs of Aquarius, Capricorn, and Sagittarius. These signs represent the systems required for the legs of the newborn person to function. These signs also lead to the appearance of the systems necessary to the body's other functions. The three signs at the ankle, knee, and hip represent the systems of the body that are responsible for body differentiation.

Aquarius, which codes the ankle and the heel, belongs to the fixed air element. It represents the neuro-vegetative system, which carries the neuronal receptors through the vast system of blood vessels. The blood vessel neuro-receptors system permeates the whole body and carries knowledge and independence.

Given that Aquarius codes the ankle and heel bone; it would seem that it allowed someone to stand on his own feet for the first time in human history. Before the Age of Aquarius, one knelt before the divine (with Capricorn representing the knee as a foundation for the human being).

Capricorn, which codes the knee, is the next sign after Aquarius. It belongs to the cardinal sign of the earth element. Capricorn represents the bones, the connective tissue, and the bone marrow. The knee represents the ability of the human being to kneel and surrender to the Divine.

The sign of Sagittarius is the next sign after Capricorn, and it regulates the next joint, the hip. This is the main joint that accomplishes the motor activity in the body. Sagittarius regulates the metabolic activity of the body through the blood vessels of the muscles, containing erythrocytes and hemoglobin.

Sagittarius also regulates the metabolism and the energetic status of the body, coding the major glands: the liver and pituitary gland. It is possible that Sagittarius also has access to regulation of the production of red blood cells. Sagittarius represents the mutable part of the fire system and participates in the burning and elimination of the products of metabolism and the products of emotional work through the reactions of oxygenation.

As a mutable sign in the fire system, Sagittarius provides the means for the human being to express emotions. By regulating the motor system and metabolism of the body, Sagittarius achieves the goals of the fire system: expression of emotions in the human body as well as in the world, as both the body and the world are homes for the human being.

Jupiter, which rules Sagittarius, is unique in its expanded structure.

FUNCTIONAL TRIOS ON THE HUMAN ECLIPTIC AND THEIR DESCRIPTION·

All three signs located on the leg represent the foundation systems necessary for the body to function on the Earth. They provide the systems necessary for nutrition, construction, and metabolism.

The self-nutrition and self-regulation begins with the sign of Aquarius, which builds the body, starting with vessels and their neuronal receptors—the vascular body. The vascular body prepares the space for the building blocks of bones and connective tissue, which are provided by Capricorn. The foundation of constructive bone tissue allows the muscle to work and the metabolism to switch on, which ignites the self-expression provided by Sagittarius.

Those three systems—the nutritional vascular system, the supportive skeletal system, and the muscular-joint motoric and metabolic system—provide the foundation for human functioning.

The independent systems that arise from the leg allow the relationship and communication systems of the body (also represented by three signs) to arise. These systems are intimacy (the constellation of Scorpio), relationships (the constellation of Libra), and the overall feeling of prosperity within the individual on multiple levels (the constellation of Virgo). The constellation of Scorpio represents the way the person connects with others on a deep level. The constellation of Libra accomplishes the multiple tasks of body regulation and connection between the nervous system and all the organs of the body. And the constellation of Virgo represents the way a person connects to and regulates his own state of being.

The qualities of the communication and relationship system of the person take the overall person's consciousness to a new level, moving it up to the part of the brain called the neocortex—which is highly developed only in humans. It represents the celestial spheres of the person, as at this area the person connects with God and the human mind connects with the divine mind.

The signs on the top of the human head represent and carry the function of the human brain, with its unconscious, super-conscious and conscious parts—the constellations of Leo, Cancer, and Gemini, respectively.

Self-expression and creative activity start with the sign of Leo. That's where the ruling sun plays such a magnificent role, inspiring all creative processes in the person.

Cancer connects the human neocortex with the divine brain through the super-conscious parts of the brain. Cancer controls, organizes, and directs the activity of the whole body.

Gemini represents the way of thinking that occurs when one thought is reflected in another, forming a conversation. This is typical for the special brain activity of the frontal cortex.

The work of the signs of the neocortex brings the development of the further group of three signs that form and carry the functions of human emotions. It's represented by signs in the front of the person's body, including the face and neck. Those are the signs of Taurus, Aries, and Pisces.

The rise of the emotional mind starts with the senses, represented by the sign of Taurus. The senses give the person his perceptions.

Next is the emotional brain of the person, which is carried by the limbic structures and represented by Aries. The emotions accumulate in the lymphatic system, which includes the thymus and lymphatic nodes and the spleen. The spleen is a place of interaction of the lymphocytes and erythrocytes, transporting the antigens connected with hemoglobin. After the interaction, the antigen is passed from the erythrocyte to the lymphocyte and can be stored as an emotional memory. Pisces regulates these interactions.

In Pisces, the qualities of the person rise to the exploration of the qualities of the collective self through their accumulation in the complexes of the lymphocytes. The induction of such a complex transforms the water into a crystallized state, initiating an organization of bigger complexes carrying a chain of memories.

The development of the collective self leads to the formation of the functional systems of the body, which are regulated by "the leg constellations." This concludes the cycle around the ecliptic.

The overall process of human well-being proceeds in its development along the ecliptic. Each year leads to new qualities, changes, and growth of the biological systems that represent the reservoirs containing and enacting the energies of the zodiac signs.

INTERACTION OF THE SIGNS ON THE HUMAN ECLIPTIC

On the level where the sign of Libra aligns with the body (the place of the fall equinox), there is a connection with the sign on the front side of the body—Aries (the place of the spring equinox). So the sign of Libra may connect there with the sign of Aries. This is the area of the main horizontal axis in the astrological chart.

The area between the equinoxes on the body is the area at the base of the brain. The base of the brain is located between Mars and the moon, ruling those constellations. This seems to be the major point of interaction between those constellations as well, with the other constellations ruled by the personal planets Mercury and Venus—the constellations of Gemini and Taurus. These interactions include those between the emotions, the senses, and the different responses and reactions to them (Diagram 2).

The connections among Mars, the moon, Venus and Mercury at the back and the front of the body allow human thoughts and perceptions of emotions, senses, and feelings to align with human deeds.

TO SUMMARIZE:

The signs that accomplish the biological design come next, one after another, and create the human ecliptic. The sequence of signs along the human ecliptic is similar to the way the ecliptic is aligned in the sky. Each sign interacts with one on the opposite side in the area of the equinoxes, where the base of the brain is located within the human ecliptic. This area connects the medulla oblongata, the pons and the cerebellum with the limbic system, the eyes, and the senses of the face.

On the human ecliptic, the personal planets that rule the signs are located around the head and face of the person (Diagram 2), while the social planets that rule the signs are located on the leg and foot of the person.

The energy of the sign regulating the system is reflected in the biological design of the body, which is in turn reflected in the human ecliptic. The positions of the planets that transfer the energy of the signs to the human body at the moment of birth construct the individual blueprint of the person's temperament, biological and psychological energies, health, and well-being.

THE REGULATIONS OF THE GALACTIC CODE:

One zodiac sign is ruled by just one particular planet of the solar system, and one planet rules one sign. The planets transmit the energy of the zodiac sign into the human body, where it is stored in a particular biological system. The human biological systems are the reservoirs of the energies of the zodiac signs. There is a human ecliptic around each person similar to the celestial ecliptic. The energies and shapes of the systems form bodily constellations that are in a way similar to the celestial constellations. The planets of the solar system form inner human planets.

The inner human planets are the energy centres of the body, which feel different aspects of the human psyche. The location of the inner planets may coincide with the location of the systems of the body they regulate. The systems of the body may have twelve centres or twelve inner planets from which they get their blueprints. Of the twelve planets, eleven (including the sun) are planets of the solar system, and one may be the fixed star Sirius. Sirius in everyone is pointing toward fifty degrees of Cancer.

The energy of the planets is similar to the energy of the signs they rule. Per this postulate, examples of the different aspects of psyche will be as follows:

Mercury – the aspect of thinking; Venus – the aspect of the senses; Moon – of the emotions; Mars – of actions; Sun – of devotion; Jupiter – of abundance; Saturn – of self-validation; Uranus – of being industrious; Neptune – of the emotional collective unconscious, dreams; Pluto – of the transformation. Chiron is the aspect of healing.

The structure of the human solar system may reflect the structure of the celestial one, with the planets arranged along the human body (Picture 7). This existence of the energies of the solar system and of the zodiac in the human body allows the microsystem to repeat the macrosystem—as above, so below.

The positions of the planets on the celestial zodiac at the moment of birth bring activation to certain places of the human systems on the human zodiac, forming a blueprint for the whole life. The blueprint of the person's energy depends on the planet that is activating a particular biological system, on the position of the system relating to the ascendant of the horizon at the moment of birth, and on the aspects and interactions of the planets on the human ecliptic.

Ultimately, the multitude of combinations of eleven planets and twelve signs, their location and their interaction yields a unique combination of energies for each person.

THE SOLAR SYSTEM AND ITS DIRECTIVES TOWARD THE EVOLUTION OF LIFE ON EARTH

THE SIGNS OF THE WATER ELEMENT AND THE BEGINNING OF THE LIFE CYCLE

The first living beings were born in water. Certain signs and planets may have a greater influence on earthbound life's beginnings. The signs of the zodiac may be taking turns in the progression of life and the development of living beings.

The water element could be the life-giver to the living being.

Cancer is responsible for the work of the stomach, which was the first organ to appear in the living cell and provide nutrients for the whole system. The stomach takes foreign material from outside the cell and transforms it into nutritional material accepted by the cell. The macrophages participate in this process by passing the antigens of the food to the lymphocytes, which form complexes with them. While the stomach changes the foreign material by one particle, the immune system works with substances that differ from the native material by one particle.

Cancer is also responsible for a part of the brain that fulfills an

41

important life function: the pineal gland, which switches the biorhythm for the day and night cycles.

Another sign that could be responsible for the beginning of life is Pisces, with its protective function for the living organism. The ruler of Pisces is Neptune, with the trans-Neptunian objects surrounding it. The trans-Neptunian planets help with the other major function of living systems: reproduction. The planet responsible for the transformation and reproductive function is Pluto.

The reproductive function of the living system starts with the process of transforming the genome and building the new genotype in the process of meiosis. The lymphocytes may participate by forming the complexes with the proliferating cells of the ovaries and the testicles. They may supply the forming spermatocytes and the eggs with the emotional or metabolic antigen molecules. The reproductive function gets more complicated with the appearance in mammals and human beings of the intrauterine pregnancy, which accepts foreign particles (antigens) for several months.

Among the water signs, Pisces seems to be the one connected with the beginning of life in the Milky Way. The collective unconscious started to come to life, giving life to living forms, and those living forms organized and protected themselves. Neptune (surrounded by the trans-Neptunian objects) carries the consciousness of the collective unconscious and has regulated the lives of the first primitive forms. Pisces may be the sign that starts the cycle of life, which starts with the organization of the body, creating boundaries to separate and protect itself. As Pluto is connected with Neptune as a trans-Neptunian object, both planets could participate in giving birth to life.

After the water signs, the regulation of life on Earth falls to the signs coming next after Pisces, in a clockwise manner.

THE INVOLVEMENT OF URANUS AND THE AIR ELEMENT

Uranus, ruler of Aquarius, is the next planet involved in the construction of living systems (after Neptune, with its trans-Neptunian objects). Uranus is the planet located next to Neptune and has an icy chemical structure rich in nitrates. Uranus also helps with the water structures or vessels that carry nutrients to all the organs and other parts of the living being. The vessels are richly supplied by the nerve endings, which carry the functions of sensitivity and recognition as well as certain stimuli to the endothelial and smooth muscle cells surrounding them, to aid in functional performance. The vessels perform the work of support and connection, which is the function of the neuro-vegetative system.

Uranus rules Aquarius, which carries the knowledge of support for the whole structural system of the living being. Aquarius belongs to the element of air.

The other signs that belong to the air element are Libra and Gemini. All air signs carry the function of connection and support.

The new and more organized life forms, which appear after the influence of Uranus, may still be living in water. Uranus prepares the work for the appearance of the new living beings that are capable of living on land and have the structure needed to do so: the musculoskeletal system, which appears with the involvement of Saturn and Jupiter.

THE INVOLVEMENT OF SATURN AND THE EARTH ELEMENT

Saturn, ruler of Capricorn, is the planet that participates next in the development of the living being. Saturn gives rise to the skeletal system of living beings and to the kingdom of vertebrates.

The appearance of vertebrates leads to the involvement of the fire system and the development of the major systems: cardio-respiratory, metabolic, hematopoietic, and reproductive. All these systems' functions are fulfilled with the participation of the planets of the solar system, starting with Sagittarius.

Unlike the icy planets, Saturn is firm and structured.

The sign of Capricorn, participating in the construction of the skeletal system, belongs to the element of earth.

THE INVOLVEMENT OF JUPITER AND THE FIRE ELEMENT

After Saturn, Jupiter is the next-closest planet to the sun. It rules the constellation of Sagittarius, which is considered to be the center of the Milky Way, to which the solar system belongs. Jupiter has the largest diameter of any planet in the solar system, though the sun is larger.

Jupiter helps vertebrates to gain the necessary energy and functions that will lead them to develop into higher vertebrates: primates and humans. It helps with the acquisition of the metabolic functions necessary to move the muscles, activate the glands, and build the hematopoietic organs needed to supply oxygen to the organs and sustain the complicated nervous system, or neocortex. This achieves control and interaction among the body parts.

The constellation of Sagittarius rules the hip, where the motor energy collects in the nuclei of the spinal cord and reorganizes in the working of the muscles. It also rules the work of the liver, with its metabolic function that provides energy for the work of the muscles. Jupiter is also responsible for the work of the pituitary gland, which controls the reproductive organs, the uterus, and thyroid function.

The other signs belonging to the fire element are Leo and Aries, which rule the brain. The constellation of Aries rules the emotional brain, or limbic system, while Leo rules the unconscious parts of the brain: the parietal, temporal and occipital lobes.

The brain, the metabolic system, and the motor system of the spinal brain carry out the functions of self-discovery, self-expression, and self-experience in the world.

Saturn and Jupiter are among the social planets of the solar system. They supply living systems with all the necessary parts and organs for self-construction and self-expression. The stage of self-expression leads to the stage of transformation and reproduction. This stage is led by the planet Pluto (ruling the sign of Scorpio as described in the signs of the water element and the beginning of the life cycle above).

THE INVOLVEMENT OF THE PERSONAL PLANETS AND THE MOON

MARS

The personal planets rule the systems characteristic of mammals and humans.

After Jupiter, Mars is the next-closest planet to the sun. It rules the constellation of Libra, as Libra is the sign of action. (In this system, the rulerships of two planets—Mars and the moon—are changed as described above.)

Libra belongs to the air element that is responsible for communication and relationships among the different parts of the living system. Libra is the part of the vegetative nervous system that is responsible for taking the lead in innervations of the body for the performance of bodily functions. Because of the twelve nerves that come from the base of the spine, Libra is sometimes referred to as "Twelve Brothers" or "Twelve Titans." Most nerves that come from the nuclei belong to the medulla oblongata, which is regulated by Libra. Libra also regulates the kidneys and the adrenals, which are responsible for the body's fight-or-flight response to different emotions.

Mars regulates the major actions of the body as well as the male hormones. Mars interacts with Venus in the construction of the throat cartilage and the vocal cords, which are different for male and female bodies. Mars and Venus also exchange the nervous fibers that control the five senses. Venus offers the senses of taste and smell, while Mars rules the fibers carrying the sense of touch. The consciousness of Mars is connected with communication and relationships of giving and receiving.

THE SUN

The sun rules Leo, which is located to the right side of the constellation of Cancer. The sun is at its zenith on the top of the chart and is at home in the sign of Leo, which constructs the unconscious world, as represented by multiple parts of the brain.

The sun represents the all-creative and all-encompassing nature of the human being. It rules the parietal, temporal, and occipital lobes of the human brain. The main function of Leo, which belongs to the fire element, is self-expression. Leo participates greatly in the creation of the neocortex and the development of the modern human being's brain. Life may need to go through many cycles of development along the signs of the ecliptic in order for the brain to reach clarity and be attuned to the Divine frequencies.

MERCURY

Mercury rules the sign of Gemini, which is located next to the left side of Cancer and in which the summer solstice occurs. Gemini, which is on top of the astrological chart, is an air element. Gemini constructs the consciousness of the frontal lobes of the brain, with their executive function, and rules the lobes as well as the rib cage and the shoulders, arms, and hands.

Gemini constructs the thyroid gland, taking upon itself the metabolism that processes food for the body in order to obtain energy for its functioning. The main purpose of Gemini, as an air element, is to connect and support. Gemini continues the work on the development of the human brain and consciousness and the relationship aspect of the brain through the frontal lobes and their function of communication.

VENUS

All signs located above the sign of Aries on the zodiac ecliptic regulate a consciousness that is above emotional consciousness on the Earth. Venus rules the sign of Taurus, which belongs to the earth element. The main function of Taurus is doing things in a right way. It rules the eyes, the face, the five senses and the throat. The main function of the throat and face is to construct and enjoy the world with the senses. Love becomes something that is designed by the person himself.

Venus is responsible for the unique thymus gland that builds lymphocytes. Lymphocytes' development goes through stages of positive and negative stimulation; this is necessary for their maturation so they can accomplish their protective function in a right way.

The consciousness of Venus is the higher consciousness of love. The human senses are preparing the matrix for the person's experiences.

THE EARTH AND THE MOON

The moon and the Earth influence the development of the human emotional mind, which is regulated by the sign of Aries (this is a new rulership assigned to the moon, based on certain criteria described earlier). Aries belongs to the element of fire.

Aries regulates the part of the brain responsible for emotions—the limbic system. Aries is located on the same axis with the constellation of Libra, and the two balance one another. The sensing of emotions is expressed in the actions of the vegetative nervous system and in the fight-or-flight response of the adrenals as well as in the production of the sexual and corticosteroid hormones.

Aries and the development of the limbic system in early childhood are strongly influenced by Libra. The relationship between parents and child and the habits and skills the child learns influence his or her emotional development.

The emotional brain is very important for the individual, and its proper development lays a foundation for the human brain and human bodily responses. The moon regulates the emotions and human responses to them. This nervous regulation balances the humoral stability on the other side of the axis, and the humoral reactions affect the neural responses. The location of the moon in the sky—its phases—regulate emotional responses in each moment.

The main function of the Aries constellation and the moon, as belonging to the fire element, is self-expression and self-recognition.

Planets belonging to the fire element are luminaries and are expansive, such as Jupiter, the sun, and the moon.

Human spirit regulates the planet Earth, which is expansive, with its colors and life visible to the human eye, and unique among all the planets. The consciousness of the Earth is connected with the development of the consciousness of the human being.

Aries and Libra are the intermediate signs located on the horizontal axis. They are connected with the mind as well as the body. The emotional brain is tightly connected with the reactions and responses of the body. They can be recognized and regulated by the person.

THE SIGN OF VIRGO

Chiron rules Virgo, which is located on the right side of Leo, in the unconscious part of the brain. The main goal of Chiron is to transform the unconscious realm into conscious perception. It is connected with the human spirit that intends to recognize and know itself. Virgo fulfills a special role in evolution: development of qualities for fulfilling the human role in its relationship to the self, others, and the world and merging with the divine.

Chiron is a centaur, the first of a class of objects orbiting between the outer planets. It is interesting that centaurs are not in stable orbits and over millions of years will eventually be removed by the gravitational perturbation of the giant planets, moving into different orbits or leaving the solar system altogether. In this way, the main essence of the person may eventually be removed from the solar system. The Chiron return measures fifty-one years, and it rules the analytical mind of the person.

The constellation of Virgo is located next to the constellation of Leo, the home of the sun.

The planet on the top of the chart that rules Cancer is located between the sun and Mercury—conscious and unconscious realms. The signs located on the right side represent the unconscious processing of the destination. The signs on the left are the ones that need to step onto the conscious path (Diagram 2).

Gemini, ruled by Mercury, signifies the higher achievement of the conscious realm. Upon maturation, the person's analytical mind, ruled by Virgo, becomes more conscious.

THE SEQUENCE OF THE PLANETS INVOLVED IN THE DEVELOPMENT OF LIVING BEINGS

The development of life on Earth follows steps connected with planets that are ruling by zodiac signs. The planets participate in the order they follow after each other, towards the Sun.

Life is first induced by the planet Neptune. After Neptune comes Uranus, then Saturn, Jupiter, and Mars. The sequence of planets in the solar system corresponds to the sequence of the development of the living beings on Earth, starting with Neptune. So the development of life on Earth follows the sequence of the planets in the solar system just as it follows the sequence of the signs of the zodiac ecliptic.

In terms of the body, this sequence moves from the toes and heels of the foot up to the knee, then to the hip. It moves upward on the back side of the body toward the back of the neck, which is ruled by Mars. The sequence then moves to the top of the head, ruled by the sun, and then descends toward Mercury on the front part of the head, Venus on the eyes and Earth and the moon on the nasal part of the face. The Earth comes into direct contact with Neptune, ruling the lymphatic system in the front part of the body, including the neck. Neptune and the trans-Neptunian objects cover the whole front of the body and descend down the legs to the toes and transfer to the heels of the feet (Diagram 2).

Here is the sequence of the planets and their sizes as they are depicted in astronomy books:

PICTURE 7. THE ORBITS OF THE PLANETS OF THE SOLAR SYSTEM

Pluto, which is one of the trans-Neptunian objects, is an exception to the sequence. Being one of the trans-Neptunian objects, it rules the water-element sign, Scorpio. Because Pluto is one of the sequences of planets connected with Neptune, it comes to stand opposite Neptune on the back side of the body between Jupiter and Mars.

Another exception to the sequence is the centaur Chiron, which was discovered only recently. Chiron locates itself between Mars and the sun, ruling part of the back brain. This location may be possible because Chiron rules an ancient part of the brain. Chiron's centaur nature means it may not belong to the solar system in particular.

The line of development follows the direction of the planets from Neptune and trans-Neptunian objects toward the sun on the back side of the body. Then it goes from the sun toward Earth on the face and again meets Neptune on the front of the body, including the neck.

Mars and Chiron are the last planets of the solar system that regulate the systems located at the back. Starting with the sun, the location of the regulated systems points to the front of the body. The locations of the person's front and back are important for the planetary regulation. The person is standing sideways looking to the left, pointing to the development of life in the clockwise direction.

If life starts as discussed with the planet Neptune and the sign of Pisces, Neptune, carrying the individual and collective emotional unconscious, gives rise to the organization of the body of the person (Diagram 2).

The other water sign, Scorpio, may also add to the supply of the collective unconscious (in the form of the reproductive system) to the consciousness of the development of human relationships. Both spheres of the collective unconscious need to go through human development to clear up and collect more light.

THE DEVELOPMENT OF CONSCIOUSNESS ALONG THE CELESTIAL ZODIAC

The sun rotates around the Milky Way in a clockwise manner (Picture 8).

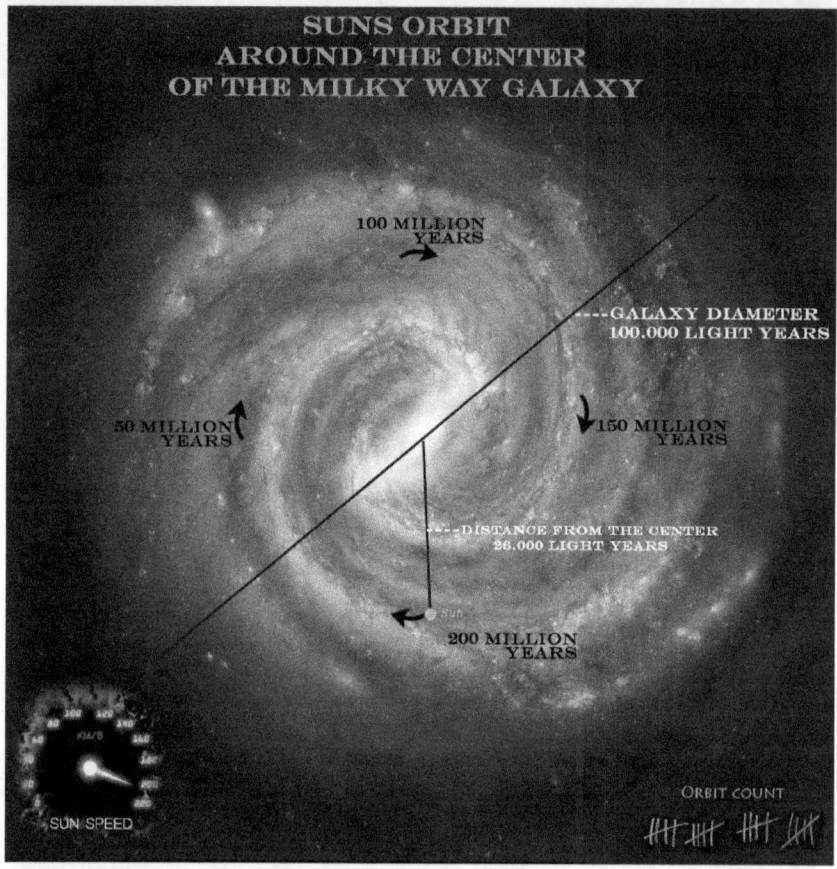

PICTURE 8· THE MILKY WAY GALAXY WITH A TRAVELING SOLAR SYSTEM

The planets of the solar system seem to regulate many processes on Earth. This happens because of their energetic connection with the signs of the zodiac, beginning with Pisces and corresponding with the signs' clockwise sequence along the ecliptic.

Many processes are connected with the influence of the zodiac signs along this sequence—the change of the Zodiacs connected with the movement of the sun, the build-up of the different biological structures along the human body, the travel of human consciousness.

Some other processes connected with the sequence of the Zodiac, such as the person's temperament, regulation of the nervous system and the stages of a person's development, will be discussed in the following chapters.

The Human Zodiac and Its Influence on Human Development

Each year, the person moves through every sign of the ecliptic at the rate of one per month. The ecliptic represents both the whole life and the main stages of the person's development, which are repeated each year through the development of his or her consciousness.

Accounting for the signs' location on the human ecliptic, it is possible to surmise that the trio of constellations involved in birth and the beginning of life is the celestial trio, or the brain trio (consisting of Leo, Cancer, and Gemini). A trio of signs represents a quarter of the ecliptic. The trio starts with Leo, which is the last sign of the ecliptic and signifies the development of the human spirit. The trio finishes with Gemini, which begins a new cycle of life.

According to this theoretical system, the sign that takes part in the organization of life and death, and darkness and light, and supplies the new spirit and clears the old accumulated spirit is Cancer. It is also the sign that organizes and takes part in the actual birth of each person.

Gemini represents the birth of a new consciousness that, in turn, becomes the source of life. Gemini is a mutable sign and symbolizes the end result of the interaction of human creativity (represented by Leo) and the celestial realm of Cancer.

Cancer clears out the human consciousness and fills the human brain with super-conscious material. The end result of this is the appearance of the cleared human consciousness birthing into life.

The movement of consciousness through the year is in a clockwise direction, starting with the beginning of super-consciousness in Cancer. It then moves to the appearance of human consciousness in Gemini, to the emotional realm, to the earthly realm, to the thoughts-and-relationships realm, and then returns to the celestial realm.

Each realm corresponds to the four "trios" of constellations. The movement of consciousness through the year in this manner is counter to the movement of the planets and the rhythm of the seasons. It resembles more the clockwise movement of the millenniums along the signs of the zodiac.

Gemini is about the beginning of human consciousness, the choices in life, and how to think about possibilities and different directions in life. It's about how to enjoy your own thoughts and bring more relationships and opportunities to life. The thoughts resemble twins traveling together.

Afterward, Gemini consciousness enters the emotional realm, represented by Taurus, Aries, and Pisces. The emotional realm starts with perception in Taurus, goes through the main structure of the emotional mind in Aries, and ends in Pisces with the collection of human emotions and their interactions. The structure and development of the realms goes clockwise, as does the development of human consciousness.

The third month of the year after Cancer is represented by Taurus. Here human consciousness arrives at the beginning of the emotional mind. The month of Taurus represents a fixed sign that establishes a template for the development of emotions. It is represented by the human senses and the throat. One's way of sensing things and talking is the foundation for the development of emotions. During this month, it's important to perceive things in a proper manner: the development of intuition and the verbalization of experience. Taurus thinks about how to perceive and do things in a proper manner. The development of the senses and the use of correct words are each person's main achievement.

DIAGRAM 3 TRAVEL OF CONSCIOUSNESS ALONG THE HUMAN ECLIPTIC

Knowing how to use one's intuition and sense everything surrounding oneself becomes the main prerogative of Taurus. The planets in each sign, as well as the transits, are the main carriers of this month's energy. The planets' aspects give important indications of the way in which a person's energy will evolve during this month.

The fourth month after birth is the month of Aries, a cardinal sign that represents the essence and the structure of the emotional brain.

Aries represents a limbic part of the brain, which collects all perceived information and manufactures and senses emotions. Anatomically, the emotional brain comes close to the nose part and spreads all way to the temporal part of the brain. It includes the memories stored in the amygdala and the hippocampus. Aries is about how to experience things—the correct development of the emotional mind is a main goal of the person.

The fifth month after birth is Pisces. Pisces represents the accumulation of emotions in the lymphatic system of generations of human beings. Pisces also represents a consciousness of compassion regarding the emotions and qualities of others. If Taurus and Aries are directed toward themselves and the process of perception and experience, Pisces is directed toward others and to helping and having compassion for the accumulated collective emotional unconscious.

The next movement of consciousness is into the trio of the earthly realm represented by Aquarius, Capricorn, and Sagittarius.

The matrix of this trio is represented by Aquarius, which forms the sensing body of neuronal axons kneaded into the structure of vessels. Capricorn represents the bones, which are the main structure of the human-realm trio. The trio ends with Sagittarius, which represents the human muscles of action in the world.

The sixth month of the year, Aquarius, represents the matrix of the creation of the human body. It is represented by the nerve endings kneaded into the body of vessels. The nerve endings represent consciousness sensing the human body. They also represent the generalized knowledge of the whole body.

Aquarius makes up the vessels that carry the essence accumulated in the prior emotional trio. They build up a human body that senses and knows itself. Aquarius constructs a matrix that allows all the other structures of the body to work properly.

The seventh month is Capricorn, which represents the main structure of the human realm: the bones and connective tissue.

These structures are the foundation of human life. The consciousness of Capricorn takes human responsibility over the main projects of life, learning to accept projects and design a life. In these ways, Capricorn imitates God or the higher power.

In the eighth month, the consciousness travels through the sign of Sagittarius, which represents the muscles of the hip and the ability to walk and expand oneself in the world. Sagittarius also represents the metabolic processes and all activities in the body that are necessary for the action and fulfilling of the life's task.

Sagittarius wants to be big and strong, like the major bodily processes it controls. It likes to travel, moving with big steps that use the major supportive joint of the body: the hip.

Next, the consciousness moves into the trio of thoughts and relationships, represented by Scorpio, Libra, and Virgo. Scorpio represents the matrix of the trio, where consciousness of relationship with another being begins. The next sign, Libra, carries the nervous structures for relationships. The ending sign that carries the results of the whole quadrant is Virgo, which establishes one's relationship with oneself.

In the ninth month after birth, consciousness travels through Scorpio, which initiates the possibility of relating to another and merging with a different being. It is the major sign that works on knowing and accepting another person. Scorpio represents the reproductive system, which focuses on similarities, differences, and compatibility between two people.

The tenth month of the year is the month of Libra. Libra is a cardinal sign that carries the structures for the relationship trio. It is represented by the nervous connections formed by the vegetative nervous system, which operates on the basis of biofeedback—the process that determines how the action of the body with its glands and nervous system affects other organs and which actions are needed to fine-tune these actions. Libra operates through the relationships between people, their interactions and effect on each other.

The eleventh month of the year belongs to Virgo, which establishes the relationship with oneself on the basis of peace, truth, health, and wellness in all areas. Virgo births the new healthy patterns of finding oneself and relating to oneself in a new and healthy way. It acts with new, built-in structures of wellness, self-esteem, and self-appreciation.

Virgo represents the final synthesis of the quadrant of thoughts and relationships and the end result of the growth and development of the earth sign.

The last quarter of this journey is through the three signs that make up the trio of the celestial realm. Leo forges a foundation for the celestial realm of the human being, creating multiple receptors for the different parts of the brain, most of which have an unconscious nature. Cancer is a cardinal sign that builds the main structures for the celestial realm. Gemini synthesizes the resulting new birth of the individual's conscious brain with its ideas, thoughts, and desires.

The twelfth month of the year is given to Leo, the fixed sign that creates the matrix for the individual's action and creativity in the celestial realm. Leo represents the biggest part of the human brain, responsible for most of the person's functional possibilities. It consists of a myriad of receptors. The consciousness of Leo is one of creation.

Owing to the incredible number of receptors and possibilities within this area of the brain, the creative potential is unlimited. This sometimes gives the individual the feeling that he needs to create more. Because there are so many choices, it's often unclear to the individual just where his greatest potential resides.

Leo is the last part of the cycle of human development—the creative work of the human spirit, which advances the maturation of the human brain.

The first month of birth is represented by Cancer. In this sign, the potential of the person and his brain enters the celestial spheres. Cancer builds on the creative potential of the brain and its receptors. It is a super-conscious part of the brain, bringing order and organization. Cancer is a cardinal sign, bringing structures that guide the top and other main parts of the brain into the celestial realm, which is capable of divine interaction with God.

Cancer is about the new super-conscious, the structuring of day and night, and the creation and beginning of life. This month is about returning to the Creator and enjoying the benefits of spirit. Cancer provides new beginnings and attends to the needs of the other signs.

The new ecliptic starts again with Gemini. Gemini represents the birth of the human spirit from the indefinite spirit, with its new goals, aspirations, and wishes. Gemini celebrates the beginning of new consciousness. It's the beginning of a new cycle of life.

THE HUMAN ZODIAC AND ITS INFLUENCE ON THE SPHERES OF THE HUMAN LIFE

This chapter repeats the information of the preceding chapter and presents the movement along the zodiac in a slightly different way.

THE GROUPS OF THE TRIOS ON THE HUMAN ECLIPTIC

Each trio on the zodiac ecliptic is composed of three signs.

The trio of constellations coding the brain and located on top of the chart (with the constellation of Cancer as a cardinal sign in the center) symbolizes the brain and may be called a celestial trio.

The three signs in front on the person's face, with Aries as a cardinal sign in the center, represent human emotions and may be called an emotional mind trio.

The trio of signs located on the leg, with Capricorn as a cardinal sign in the center, symbolizes life on Earth and may be called an earthly trio.

The trio on the back side of the person's head, with Libra as a cardinal sign in the center, represents the regulation of human thoughts and relations and may be called a relationship mind trio.

Each trio starts with a fixed sign representing the receptors and the foundation template for the trio. The activity of the receptors produces the matrix for the trio's function.

The trio's center is occupied by the cardinal sign that represents the essential structures of the trio, the source of accumulating and storage of the element that it represents (described more below in Chapter 1).

DIAGRAM 4 INTERACTION OF THE SIGNS IN THE ZODIAC

The trio ends with the mutable sign, which brings the final effect: the function and elimination source for the whole system.

The celestial trio starts with the sign of Leo, representing the brain receptors for skills that need more research. As is always the case in the fire-element systems, the element that controls the function and serves as a transporter for the receptors and functioning molecules is oxygen. The functioning cells are probably the Purkinje cells.

Cancer represents the structural cardinal sign of the celestial trio. Functionally, it brings the super-conscious element that separates the unconscious and conscious parts of the brain. Cancer represents the entrance of another consciousness, different from our Milky Way. Cancer's function may be clearing the consciousness that is building up in the solar system and bringing a different consciousness (one from another galaxy).

The emotions and consciousness of the solar system, concentrated in the sign of Leo, is cleared in this way—merged with the celestial essence for one and moved toward the creation of the new human being. Gemini starts the new life cycle, bringing the consciousness and the help of the frontal lobes and their executive functions.

The emotional mind trio in the front of the person starts with the fixed sign of Taurus, which represents the human senses: eyesight, hearing, smell, taste and the sensation of touch on the face and throat. Taurus is considered to regulate the eyesight, throat and human speech.

Human senses and their work, as well as words, are the template of the emotional mind of the person. Aries represents the structure and the essence of the energy of the emotional mind.

Aries is a cardinal sign, located in the center of the emotional trio of signs. It represents the limbic system of the mind, which is sensitive to human senses and words and reacts to them with a set of emotions.

The trio ends with the mutable sign Pisces, which represents the individual and collective emotions of humankind, bringing the result of the emotional state of each person to the collective state.

Physiologically, Pisces represents the human lymphatic system. Emotions accumulate in the lymphocytes, the thymus, lymphatic nodes and vessels. It's interesting that lymphocytes are produced by the thymus gland only until age thirteen or so, marking the end of personality-building and the beginning of sexual maturation. In those thirteen years, the person accumulates the individual emotions that are coded into the immune system. After this age, the thymus undergoes regression and atrophy.

The collective emotions are directed into the systems responsible for physical health. These systems are regulated by the signs of the earthly trio. The emotions clear themselves out during the whole human life cycle, which is represented by the ecliptic of the zodiac.

With each human life, the collective emotional unconscious shows up in the life and changing. And by showing up, it clears itself from the consciousness of the solar system. In the sign of Cancer, it may find an exit into another galaxy. The collective unconscious is further cleared with each individual life.

The earthly trio starts with the sign of Aquarius, which pours the consciousness of the lymphocytes and the emotions into the blood vessels. Aquarius starts the physical life on Earth by pouring consciousness into the person and building the system of blood vessels. The blood-vessel network represents the template of the vascular body, with its nerve endings, spreading self-regulation all around the body.

The vessel is a separate body that creates a template for the whole physical body. It is a bio-template for the connective tissue and the musculoskeletal system. The blood vessels are connected to the bones by the bone marrow and connective tissue, which are represented by Capricorn—the cardinal earth sign representing the major essence of the earthly trio.

The result of this integration shows up in the work of Sagittarius. This sign is responsible for the metabolism of earthly life, with its oxygenation and motoric activity. Sagittarius is responsible for the big muscles of the hip and for walking.

The fixed sign of Aquarius represents the receptors of the body and serves as a matrix for the signs of the earthly trio, the bones and bone marrow, and the muscles. The trio finishes with the mutable sign of Sagittarius creating metabolism and the physical activity of life on Earth.

Sagittarius, with its adjustment to the earthly plane, is deeply connected with Capricorn, which carries the structures of the human brain: the bone marrow. The bone marrow produces the erythrocytes that participate in oxygenation and metabolism. Their connection is indicated by both signs' being connected in the center of the galaxy—at zero degrees of Capricorn.

The signs of the relationship mind trio represent the next quarter of the chart. They regulate the systems located at the back of the person's head.

The relationship mind trio starts with the fixed sign Scorpio, representing intimate relationships and the mystery of merging with another in the reproductive cycle. The intimate connection of Scorpio serves as a template for the other signs of the relationship mind trio.

Libra is the cardinal sign signifying the essence of relationships and their regulations. Libra represents the vegetative nervous system, which has relationships with the body parts and regulates the internal organs by sending electromagnetic currents.

Libra regulates its relationship function in correlation with the main law of the body: biofeedback. The body answers in response to the organ it regulates, accelerating the positive response and slowing down the negative. Libra regulates the biological interactions of the body and its internal systems in accordance with the major law regulating psychological energy as well.

Libra—the scales—balances the responses for the body systems and the relationships between people. It builds personal relationships that are directed both toward oneself and toward another person, based on the principle of biofeedback and regulating and balancing the person's emotions and thoughts.

The relational trio ends with the mutable sign Virgo, which represents the state of personal psychological well-being—a mastery of feeling good about oneself. It represents the integration of the work of the relationship mind trio. At the end of the integration comes solid self-esteem and well-being within all major areas of human existence.

Thus, the fixed sign Scorpio builds the matrix of the intimacy in the relationship mind trio, the cardinal sign Libra creates the structure of the relationships, and the mutable sign Virgo brings the final product of the interaction: the creation of a relationship with the self.

GROWTH AND DEVELOPMENT ALONG THE TRIOS OF THE ZODIAC

The consciousness of the person begins after the interactions of the higher mind, where the emotions of the sun and the solar system in the sign of Leo approach the consciousness of the structures of the higher spheres of the brain in the sign of Cancer. The consciousness that arises in connection with the frontal lobes is regulated by Gemini as part of the executive function of the human brain.

The arising consciousness starts moving through the senses and the emotional mind of the person. The senses form the foundation for the emotional mind. As the senses of the person are influenced and activated in a certain manner, the limbic system of the person follows the same pattern.

The flow of emotions influences the generations. The emotional mind of the person generates the individual and collective consciousness that, in its own way, influences each human being during his development. The generational emotions flow through the life cycle of each human being to be cleared, purified, and released.

In the next step that consciousness moves through, the emotions enter the physical realm and the physical body is created. It starts with the creation of the body of knowledge—the vascular body with its nervous receptors. This gets dressed in the physical body of bones, connective tissue, and muscles.

The consciousness of the physical body is in the energy of the muscles and the metabolism, expanding and expressing itself in the world. The physical body starts relationships with others, which brings the person into the wisdom of the relationship mind realm. This conveys the person back into the celestial realm of higher mind, which represents the unconscious, conscious, and super-conscious parts of the brain.

The relationship mind goes from the stage of intimacy and

transformation to the relatedness to oneself and others and on to further connection with oneself. The consciousness of relationship mind directs the relationships with oneself toward worth, content, and prosperity.

The new emotions, human achievements, and relationships stream to crystallize in the new consciousness of the sun and then enter the signs of the celestial spheres that regulate the interaction of the human and divine minds. The merging of human and divine, in turn, gives birth to the new human consciousness and leads to the next cycle.

Human consciousness will undergo further cleansing and transformation before it goes on to the next round of development—ready to give birth to a new consciousness and a new round of life for the human being.

THE INFLUENCE OF THE INNER SOLAR SYSTEM ON HUMAN DEVELOPMENTAL MILESTONES

As described in Chapter 4, the person may have the inner structure of the solar system, which may be the essence of all aspects of human consciousness, or the cognitive system.

The solar system may have a location within the body, with the planets situated in the systems they represent. The planets of the solar system may create the little homunculus inside the body where the centers might still be rotating around the vertical axis introduced by the sun. The distance between the planets in the human solar system may reflect the distances between the planets in the celestial solar system.

The planets are located along one elongated axis in the person's body (Diagram 2). This way, everything that's happening with the planets may have importance to human development.

The consciousness of the planets is similar to the consciousness of the signs they rule. Per this postulate, examples of the different aspects of consciousness as it was described earlier are as follows:

Mercury - the aspect of thinking; Venus – the aspect of the senses; the Moon – of the emotions; Mars – of the actions; the Sun – of devotion;

Jupiter – of abundance; Saturn – of self-validation; Uranus – of being industrious; Neptune – of the individual and collective emotional unconscious, and dreams; Pluto - of the transformation. Chiron is the aspect of the consciousness of healing.

Each planet's time of rotation around the sun has an influence on the development of human systems.

THE PERSONAL PLANETS

Observing the rotation times for the personal planets, we can presuppose that they reflect the stages of human psychosocial development. Each planet reflects the development of the system it regulates in the process of making a circle around the sun. The development proceeds like this:

- The moon connects with the development of the human emotional consciousness at one month.
- Mercury influences the development of the central nervous system and the appearance of the sensory-motor stage of development at around three month of age.
- Venus influences the appearance of the sensory system, the stranger anxiety, at around eight months of age.
- Earth influences the individualization of human consciousness at around one year of age with the beginning of walking.
- The sun influences the development of the human brain and the gradual acquisition of learning skills.
- Mars influences the stage of autonomy at around twenty-two months.

    The personal planets are located at the back and front of the face.

THE SOCIAL PLANETS

The rotation of the social planets also points out the milestones of the person's later development.

Jupiter is responsible for the consciousness of the motor system, and possibly for the overall metabolic and general human development. Jupiter moves through one sign of the Zodiac each year and in twelve years accomplishes its rotation around the sun. By that time, the child may be psychologically ready to enter the world of the adult. Jupiter may also accomplish the overall development of the human muscular system, including gross and fine motor skills.

Saturn gives the person self-validation and the ability to appreciate and rely on the self. This stems from the durability of the bones, which are responsible for supporting the person through all his activities. Saturn rotates around the sun in 28.5 years and finalizes the person's psychosocial development at that time.

Jupiter is located at the area of the hip and represents large muscles. Saturn is located along the knee and represents the highly organized structures of the bone, which teach the person structure, discipline, order and satisfaction.

THE OUTER PLANETS: URANUS, NEPTUNE, AND PLUTO

Through the duration of their stay in particular signs, the outer planets facilitate the achievement of certain developmental milestones.

Uranus stays in each sign for roughly seven years, the age at which the child identifies with the parent of the same gender. This marks the beginning of an industrial phase in the child's development of gender-reflected games and activities. The child plays out and symbolically reflects his role and achievements in life. At this time, the child starts to find his position in life psychosocially.

Neptune remains in one sign for roughly thirteen years. As pointed out earlier, at thirteen years of age and the start of sexual maturation, the thymus (the organ that develops lymphocytes) stops working. As the lymphocytes accumulate new emotions, this is the time to collect those emotions and store them.

The biggest influence on the emotional life of the person happens before the age of thirteen. The collected emotions will influence the person's emotional storage for his whole life and will structure his personality. So at this age, sufficient emotional abundance must be stored to last a lifetime. By age fourteen, the personality is formed and in case of any abnormalities may be classified in terms of psychiatric classification.

Pluto (one of the trans-Neptunian objects) teaches the transition from individual to collective consciousness of having a family. Pluto remains in each sign for roughly twenty-one years, at which point the person's consciousness transitions to a collective one. The person matures to the degree that he can take responsibility for another person, for a family, for a collective life. This is a revolutionary transformation in the person's own life and consciousness.

THE DESCRIPTION OF THE DEVELOPMENTAL MILESTONES REGULATED BY THE INNER PLANETS

The inner planets are located around the head and face. They determine the growth of the brain systems that develop during childhood.

As the outer planets mark developmental milestones of adolescents and adults, so the inner planets mark the developmental milestones for developing emotions and communication skills of early childhood.

The time of the return of the personal planets relates to the stages of the child's personal development and the biological transformations of his nervous system. It's well-known that the major stages of child development occur at ages one month, three months, eight months, one year, and approximately two years, corresponding respectively with the return of the moon, Mercury, Venus, Earth, and Mars.

At one month, the child starts experiencing emotions toward other

71

people and focuses on the faces and smiles of the parents. The emotional consciousness arises.

When the child is three months old, the primitive reflexes disappear. The maturation of the nerve synapses leads to improvement of the child's normal activities and the appearance of the normal neural response. The child starts to hold up his head and prepares to explore and communicate with the world. Mercury starts the sensory-motor stage of the person's development and begins the maturation of the central nervous system (CNS). Like other air planets, Mercury helps to raise the neuro-regulatory consciousness of the person.

At six months, the child starts to separate himself from the mother and develop his own senses. The age of eight months is marked by the appearance of the stranger-anxiety stage as the child establishes his vision of the world and feelings toward his surroundings. This is the beginning of separation or individualization. The child starts to distinguish known from unknown, familiar from foreign, loved from unloved.

At around eight months, Venus triggers the development of senses and the rise of self-consciousness, which relates to informational processes, as the eyes relate to information.

At around one year of age, the child climbs to his feet and starts walking. This marks the end of the individualization phase. The child learns to manage his anxiety, to sense known and unknown, and to be emotionally independent.

The influence of the Earth marks this appearance of the independent emotional consciousness. As a result, the child learns to take care of himself and his environment. The emotional consciousness is about self-knowledge, creativity, metabolism, and, later, the heights of the human spirit. The emotional consciousness is probably connected with the metabolism of oxygen and the development of the smooth muscles.

The emotions begin in the limbic system of the brain and the products in the conjunction with the lymphocytes accumulate in the thymus, which produces lymphocytes until the age of thirteen. This is reflected by the movements of the water planet Neptune.

Neptune regulates the accumulation of emotions and gives rise to the individual and collective emotional maturity consciousness, which is guided by the water planets. As some emotions are usually hiding inside, the individual unconscious is responsible for the development of personality, which is usually checked at fourteen years of age. The individual unconscious and the collective unconscious provide the material for further development of the human lineage.

At around two years of age, the child begins the major movement toward recognition and understanding relationships. He re-connects with the mother as the major provider and starts to identify with surrounding people.

Mars, another planet of communications and neuro-regulation consciousness, begins the stage of autonomy and awakens the ability to switch engagement in independent activities and re-connect with adults.

The return of the moon signals the beginning of emotional connection, the return of Mercury the beginning of communications, the return Venus the beginning of sensory perception, and the return of Mars the beginning of relationships with oneself and others.

The moon, the Earth and the other fire planets begin the cycle of individual emotional development. Jupiter continues this development, including the emotional development of the muscles. The sun aids in the emotional development of the brain.

Venus and the planets of the earth element start the cycle of maturation for the person's senses and reliance on oneself, or self-consciousness. Saturn adds reliance on oneself and one's own psychosocial development, Chiron reliance on the maturation of one's brain and self-trust.

Neptune and Pluto, both planets of the water element, aid in the maturation of adolescents into adults. Neptune regulates the development of the adolescent's emotional consciousness from individual to collective. Pluto regulates the emotional maturity of the adult, changing individual consciousness into collective community consciousness.

Sirius acts like a switch from the highest achievements of the sun's individual consciousness to the development of the new collective seed of consciousness in Mercury.

THE ROLE OF THE NERVOUS SYSTEM

Neuro-regulation teaches the person to find his excellence in life and develop from an individual way of thinking into the community way. Mercury, Mars, and Uranus—the air planets—regulate the work of the nervous system and the cycle of neuro-activity and the neuro-regulatory connections. Uranus adds neuro-development at the latent stage and identification with the same gender. The air planets teach the person how to relate to society and develop the consciousness in a collective way. They present the stages of the development of the nervous system, which are helpful to assimilate and make more mature throughout the entire life. The stages consist of the work on the thoughts, on the relationships and mirroring of certain people's images.

Regulating the thoughts teaches one to control the senses. Mindfulness, cognitive therapies, and breathing are developed here.

Regulating identification and role modeling teaches one to build psychosocial development, which is necessary for life. Role modeling and identification with the parent, the therapist, and other significant people in one's life occurs here.

Regulating one's sense of autonomy and feeling gratitude for a supportive environment teaches one to build stability and wellness of mind.

HUMAN TEMPERAMENTS AND THEIR CONNECTION WITH THE ELEMENTS

THE QUALITIES OF THE ELEMENTS

After much questioning of the origin of the four systems of the body, I came to the conclusion that they are regulated by the four elements of the nature. The left side of the zodiac is represented by the elements that are carrying an individual consciousness, the right side by the individual consciousness.

The element of water on the top and the element of earth at the bottom represent the elements that can change their states. The horizontal axis holds the elements that oppose each other in this way. The oxygen on the left side controls the fire metabolic system; it has one state and represents the individual consciousness. The charged cations, or the electromagnetic field, on the right side controls the air element. It is represented by the cerebrospinal fluid that holds the cations in storage. It has a collective consciousness of the wave.

The water element on the top has both states, the crystallized water on the right side and the simple state of the water on the left side. The lymphocytes of other cells of the immune system accomplish the transition from the simple state to the crystallized state by forming the complexes with different substances, including antigens.

The earth element at the bottom of the Zodiac is controlled by the light that is stored in the bones. The light has both states. On the left side of the zodiac is the light in the form of photons, which is transformed into the light in the form of waves on the right side. The cells that have the capacity to transform it are from the three systems of the earth: the osteoclasts of the bones, the photoreceptor cells of the eyes and the Purkinje cells of the cerebellum.

The cardinal signs of the elements have the function of storing the element and distributing it. Water is stored in the stomach, oxygen in the nose and the nasal and frontal cavities, light in the canals of the bones, and electromagnetic waves in the cations and anions of the cerebro-spinal fluid.

The qualities of the earth element, which has both wave and particle controls, combines the individual and collective consciousness. Those are determined by the light. The same is true of the water element, with its two liquid states of being; it combines individual and collective consciousness.

The qualities of the fire element are always individual. Oxygen determines the metabolism and creativity of the body.

The qualities of the air element are always collective.

As the elements of the zodiac follow each other, the earth consciousness of the visual systems is followed by the individual emotional fire consciousness of the Milky Way. This is followed by water consciousness of the antigen lymphatic systems, which is switched to the neuro-regulatory system of the air signs.

In this cycle, the earth element determined by the photon of light stimulates emotional element, determined by oxygen in the body. The oxygen stimulates the simple water to enter the crystallized state. Crystallized water activates the electromagnetic waves of the nervous system. And the electromagnetic waves stimulate the waves of light to transform into photons.

PEOPLE'S TEMPERAMENTS AND THEIR CONNECTION WITH EACH OTHER

People born in different signs of the zodiac have different experiences of growth and transformation, depending on the element to which their sign belongs.

The experience of the person goes along with the position of this element on the Zodiac ecliptic (Picture 1). As the development follows the zodiac in a clockwise direction, each person leans toward the element that follows and tries to connect himself with the element that comes before him. At the same time, the person expresses the type of thinking of the element he's in.

People who belong to the earth element have a temperament controlled by light and may be able to switch between the individual and collective states of consciousness as well. Their thinking is impulsive. Such people are grounded by the nervous system of the air element that precedes them. The nervous element carries the consciousness of communication and love for community.

The individual spirit requires the person to the independent state of creativity and expressing himself. People belonging to the earth element love independence and are expressing themselves; on the other hand, they want to socialize and express community spirit. Those who do not achieve the merging of the two consciousnesses exhibit a melancholic temper. In normal circumstances, the person combines love of community with his individual state of creativity and expressing himself.

As the water element can switch the individual emotional consciousness into the collective, the person in the earth element has to achieve the opposite transformation. He changes the collective community consciousness into the consciousness of independence and the work of spirit.

The transformation of one consciousness into another is unique for the water and earth elements.

A person whose sun belongs to the water element has a bubbly and generous way of thinking based on the element of the water. He tries to connect himself with the individual consciousness of creativity. These people love the communicative and logical way of living and merge it with their emotional and creative consciousness. If the person doesn't combine those well, he exhibits a phlegmatic temperament. On a regular basis, these people exhibit the collective community nature with a caring, friendly attitude and a generosity toward others.

The conflict experienced by people belonging to the earth and water elements is between the collective-community and private states of being, between the community state of living and the ability to creatively express themselves.

People belonging to the fire element always have an independent, individual consciousness. In normal life, they have a passionate and expressive way of thinking. They are generous and have a warm, giving way of living. The individual quality of water is expressed in their desire to do it for themselves. They ground themselves in the examples and mirroring of the teachers or other inspiring people in their lives. If they don't achieve this, they exhibit a sanguine temperament.

A person belonging to the air element is a logical thinker and has a strong mind. He likes to exhibit his teaching capacities and discuss ways to help people in general. He has the generosity, warmth, and giving way of life that are in the community spirit, exhibiting collective consciousness of the water element.

If the person doesn't achieve these qualities, he exhibits a choleric temper.

Humankind is placed in the energies of the Milky Way to work out the consciousnesses of all of the four elements.

Interactions Between the Ruling Planets Within One Element

The Location of the Ruling Planets

The human inner planets that transfer the energies of the signs also make triangles on the human ecliptic similar to the systems they regulate (Pictures 2A-5A).

The Planets of the Water Triangle

The water planets regulate the human systems that have mostly humoral and water regulations. The humoral regulation is important because the preceding element is the fire element. The oxygen stimulates the simple water to transform to crystallized water.

Neptune, along with the trans-Neptunian objects, and Pluto represent the individual and the collective consciousness and also act as a switch from individual to collective consciousness. The human systems born in water signs act like transmitters, from the individual emotional consciousness of the fire system to the collective consciousness of the air system.

Neptune is located at the end of the sequence of planets, which starts with the sun on top of the head and goes in two directions (Diagram 2). In one direction, planets descend through the front of the face in a regular order: Mercury, Venus, and Earth with the moon. In the other, planets descend through the back of the head and body: Chiron, Mars, Pluto, Jupiter, Saturn, Uranus, and Neptune.

As is the centaur's nature, Chiron comes into the sequence *out* of sequence, as it is usually located between Saturn and Uranus. Pluto comes through its location near Neptune, on the back side of the body.

Neptune regulates the person's heels and lymphatic vessels. It holds the personal energy that is in front of the body. In this way, Neptune and trans-Neptunian bodies cover a big space around the person's body.

The emotions descend from human emotional consciousness on Earth and run to the space of Neptune and the collective unconscious. From there, Uranus pours them into the vessels of the human being. Thus the tail of the solar system is coiled into the planetary ecliptic to meet the planets that are starting the sequence at the frontal plane of the person's body on the human ecliptic.

Pluto is located close to Neptune. The space between the two planets pierces the body and Pluto locates on the back of the body and the back plane of the ecliptic. It's possible that the whole Kuiper belt energy is located on the front of the body and pierces the body.

The water-element energy permeates the head, pierces the body and envelops person's sides and feet.

The asteroid belt would be located at the back of the body between Jupiter and Mars, executing a regulatory function.

THE PLANETS OF THE EARTH TRIANGLE

The earth planets are arranged along the same axes as the water signs, which they oppose (Picture 6). The systems regulated by the earthly planets use light regulation, which follows strict rules and defines itself with codes of right and wrong. The coding of the earth systems may remind some of computer coding.

The planets regulating the earth signs are structured—Saturn, Venus and Chiron. Saturn requires discipline and organization; Venus requires recognition of right and wrong and doing things in a right way; and Chiron requires the analysis of things in making decisions about right and wrong.

The earth signs' structure programs them to choose between two options. For Capricorn, the choice is between validating and not validating oneself. For Taurus, it's between behaving in a right or a wrong way. For Virgo, it's between making a right or a wrong decision.

The earth signs need to study, learn, grow, and improve. They need to have "right" definitions and achieve results. The earth planets teach self-consciousness and learn to study themselves, getting further along in the process of maturation with each life cycle. Saturn teaches the maturation of psychosocial development, Venus the ability to recognize things with the senses to and love things more, Chiron the maturation of the mind and the ability to analyze and appreciate things more.

Saturn is larger than other earth planets and brings self-consciousness, self-reliance, and self-validation. Venus is about the size of the Earth and carries the personal self-consciousness of the face and its senses. Chiron carries the self-consciousness and maturation of the brain.

The element of earth is constantly developing along its three planets. Saturn rules the cardinal sign Capricorn, Venus the fixed sign Taurus, and Chiron the mutable sign Virgo.

Saturn is a planet with a complicated structure that achieves a necessary regulation of the main supportive structure of the human body—the bones. Saturn teaches how to support the whole body. Venus teaches control of the senses and how to distinguish right from wrong. Chiron is a dwarf planet with comet-like behavior. It's called a "wounded healer" and heals its own consciousness. Chiron regulates the human brain and learns to do the work of self-growth and appreciation.

Both Venus and Chiron do the work of love and appreciation, while

Saturn is more structured and continues the consciousness of the bones to improve self-reliance.

As Capricorn regulates the bone, the bone may rely or not rely on oneself. As the location of the sun speaks about devotion, a person with his sun in Capricorn wants to validate himself. Such people become upset when circumstances do not allow this.

Taurus regulates the senses. People with their sun in this sign distinguish good from bad and right from wrong. They prefer to meet circumstances that allow them to do things the right way. They get upset when this isn't the case and things go wrong.

Virgo regulates the human ability to make decisions. A person with his sun in Virgo likes to analyze things in order to make a good decision. He gets upset if the decision turns out not to be right.

I assume that light has the most influence in these three systems. Light takes a wave or particle form, so these systems exhibit an individual or collective consciousness. The person is more individual and struggles for his community consciousness.

Both the earth- and the water-born person go through the conflict of the cardinal horizontal axis—between individual and collective consciousness—in order to resolve and embrace. The earth-born person does it in a direction from the collective to individual, the water-born person in a direction from individual to collective.

THE PLANETS OF THE FIRE TRIANGLE

The consciousness of the fire systems is the consciousness of the Milky Way with the constellation located at its center— Sagittarius and the biggest planet, Jupiter. It is the consciousness of the sun as well as of the only live planet in the solar system, Earth with its moon.

The systems regulated by the fire planets possess emotional consciousness, the regulation of which is the biggest mystery of our galaxy. In contrast with the water systems, the systems regulated by the fire planets bear an individual consciousness. Earth and the moon carry the emotional consciousness of the human being, with his ability to care for himself and for his planet. The sun carries the emotional consciousness of the brain and its ability to create. Jupiter carries the emotional consciousness of the body. The regulation of the emotional joy of the muscles and the emotional consciousness of abundance, expansion, and happiness of the body is still a mystery.

The planets of the fire system travel in a regular rhythm through the signs of the zodiac, giving the person a regular experience of all twelve signs. On the physical plane, Earth provides an experience of the seasons, Jupiter of the years, the sun of millennia.

THE PLANETS OF THE AIR TRIANGLE

The planets of the air systems have neuro-regulation. They supply the systems they regulate with a collective consciousness.

Air-element planets stand opposite the fire signs on the axes of the human ecliptic (Picture 6). In contrast with the fire signs, they possess a collective consciousness.

The planets belonging to the air element are small—Mercury, Mars, and Uranus. Uranus is bigger than the other two, as it relates tightly to the larger social planets. It builds the matrix of the human body.

SIGNS AND PLANETS WITH COLLECTIVE AND INDIVIDUAL CONSCIOUSNESS

As described above, the water and earth signs combine individual and collective consciousness, the fire signs are of individual consciousness, and the air signs of collective consciousness.

People whose sun is in a water sign may switch between individual and collective consciousness. They need to unite the two within themselves and struggle for privacy. People whose sun is in an air sign have a collective consciousness and need to communicate. People whose sun is in a fire sign have an individual consciousness and a private nature.

I. CORRELATION OF PEOPLE'S TEMPERAMENTS WITH THE OPPOSITIONS OF THE WATER-EARTH AXES

The three axes of the earth-water signs show the opposition of the analogous and digital consciousness of the earth element to the collective and individual consciousness of the water.

The water changes its state from a simple to a crystallized form. The regulation of the earth element is through light, changing states from a photon to a light wave. The earth element contains structures having long chains that influence the transformation of the light. The purpose of the signs of the earth element is to make a right choice, a right decision, to take responsibility.

The structures of the water element form complexes that form crystallized water. The purpose of the water is to multiply and embrace all the foreign material, making it its own.

Because of differences in consciousness, the signs on the axes pull in different directions. The consciousness of water pulls into the warmth and generosity of the collective. The consciousness of the person pulls into standing for oneself and learning to achieve things in a better way.

The planets that rule the water, Neptune and Pluto, also carry both the individual and collective consciousness as well as the signs of Cancer and Sirius.

Neptune is about changing the collection of the individual emotional consciousness present at the start of puberty into a collective consciousness. Pluto is about the maturity to enter the balanced individual and collective consciousness and the person's readiness to start a family, which represents a revolutionary transformation for the genotype and the adult mind-set. Cancer is about the maturity of combining the individual and collective consciousness and relating to family, friends and oneself.

The front of the body carries more of an individual charge, so the individual consciousness prevails. At the back of the body, the collective consciousness prevails.

Cancer and Capricorn are two signs where the divisions of the individual and collective are balanced. Those signs are transitional and balanced in terms of the dominance of the individual and the collective in the temperament of the person.

The water-element person is a community person, but at the same time he likes to have private time to exercise his creative abilities. Meanwhile, the earth element has its own light mechanism of regulation and is designed to do the same task of shifting consciousness. The earth-element person enjoys individual time and the ability to work on achievements of spirit at the same time he likes to feel himself in the community and experience collective consciousness.

The planets ruling the earth—Saturn, Venus, and Chiron—have one ability at a time, and the task of the human being is to be able to switch from the collective human consciousness that he is acquiring with his development toward the heights of the human spirit, with individual consciousness.

The ability to achieve this independent switch makes earth consciousness similar to water consciousness. As the movement of consciousness on the zodiac goes clockwise, in the water sign the consciousness switches from the individual to the collective—from the fire element to the air element. The collective consciousness prevails; the person works on his private spirit and expresses himself.

In the earth element the movement goes from the nervous system of the air element to the fire element, from the collective to the individual. The individual consciousness prevails; the person works on remaining in the community consciousness.

Because of their unique regulation, the earth planets feel good in the signs they rule but may have limitations when placed on the opposite side of the axis. They may be critical or behave for the benefit one's own state of mind.

The behavior of the person throughout his growth is required to move toward being collective, and the person achieves this through the maturation and development of his nervous system.

The zodiac shows that there are three major ages and ways of working with the nervous system. This happens with the natural regulation of the signs by the nervous system and will be described in the following chapters.

2. CORRELATIONS OF PEOPLE'S TEMPERAMENTS ACCORDING TO THE OPPOSITION OF THE AIR-FIRE AXES

The fire and air signs pull in opposite directions. The fire signs are about individual self-expressive, passionate, and creative ways of doing things; the air signs are about the communicative, collective, and explanatory ways of doing things.

The regulation of the signs is humeral for the fire system and nervous for the air system. The help of neurotransmitters expresses the mystical nature of emotional consciousness.

In fire-element people, the consciousness moves clockwise from the earth element to the water element. The temperament always remains individual, as the element is oxygen, and oxygen always remains as a substance as opposed to a wave. It determines the consciousness of the front of the body, where the oxygen of the emotional system prevails. The water and the light of the cardinal signs in the solstice points also remain in their individual substance state and connect to the oxygen in those states.

The moon and the Earth express the person's creative emotional process of finding himself. The Earth allows the individual personal consciousness to take care of himself and of the planet. Meanwhile, the sun expresses the creative emotional consciousness of the mind's activity, and Jupiter expresses the creative emotional consciousness of the muscle activity.

The fire-elements people are strong in their passionate feeling. They run on fire, on the burning capacity of the oxygen. In air-element people, the temperament remains collective all through their lives. The consciousness moves from water to earth, which are both turned to the air element by their collective state, in the state of crystallized water and the state of a wave of light. The collective state remains the same, as the element of the air element is the wind or the electromagnetic waves of the back of the body.

The planets of the air sign—Mercury, Mars, and Uranus—have a collective consciousness.

Mercury develops the thoughts from the frontal part of the brain, which employs a collective form of communication with others; the thoughts might as well communicate among themselves. Mars is about integrating the actions of human mind and body. And Uranus is about the inner person's brain for the whole organism: the functional neuron receptors. They are distributed throughout the body to regulate things, communicate with each other, and regulate various functions.

The signs of the water and earth elements present ways of being here on the planet Earth, while the signs of air and fire elements present ways to do things. In this way, the energies of these two pairs square.

THE REGULATION OF THE HUMAN SYSTEMS BY THE NATURAL ELEMENTS

The major systems of the body representing elements of the zodiac are themselves regulated by the elements. The systems are fire (the emotional system), water (the antigen system), air (the nervous system), and earth (the visual representation-bone-cerebellum system).

The elements regulating human systems in a clockwise manner are:

- Water – for the antigen system.
- Wind or electromagnetic waves – for the nervous system.
- Light – for the earth system
- Oxygen – for the fire system.

Those elements transmit information and control the systems, serving as the medium for the systems to function.

The cardinal sign is the one that accumulates the element, the fixed sign is the one that uses the element, and the mutual sign is the one that dissimilates the element. It's important to note that all four human systems still accomplish their other functions, thus maintaining physical well-being.

1. Water serves in two forms: simple and crystallized. Crystallized water conducts the H+ proton's waves. Water makes the transition from the individual form of the element to the collective, while the earth element makes an opposite transition, from the collective wave form to the photon individual form. In each water sign, both forms exist and transform from one into the other. The movement of the H+ induces the water waves.

Water serves as a transmitter for the water-antigen system. Antigens are carried in the system by the lymphocytes—glial cells—so the lymphocytes participate in the induction of the crystallized water. Induction may be achieved by the lymphokines, which induce water to shift from the simple to the crystallized form.

Cancer, as the cardinal sign, holds the water; the macrophages transport antigens from the stomach by forming complexes with the antigens and using crystallized water. Scorpio uses the lymphocytes (or the other cells belonging to the immune system) to transport antigens to the proliferating reproductive cells—egg and spermatocytes—forming multiple forms of the receptors and possibly influencing the genotype.

The immune system may use the antigens of the preceding fire emotional system to build the antigen complexes. Consequently, the emotions that build up through the sexual system (as well as other emotional and metabolic processes) may play a role in the antigen complexes that are built up and presented on the reproductive cells, helping the egg and the spermatocyte to recognize each other and so facilitate reproduction.

Pisces uses lymphocytes for the elimination of antigens that are pathogenic for the body. Lymphocytes are also used to make the complexes and protect the antigens from elimination, widening and multiplying the antigen composition of the body. The thymus may employ the interaction of the lymphocytes with the macrophages for the selection of antigens to be eliminated.

2. Electromagnetic waves of the body, or the wind, make up the transport and control element of the nervous system. The wind waves exist only in the collective form of waves. The wind is composed of electromagnetic waves, between the cations and the anions. The wind may be also accomplished through thought waves.

Libra, as the cardinal sign, accumulates electromagnetic waves and distributes them toward the nervous receptors. Libra stores electromagnetic waves in the form of the cerebro-spinal fluid of the spinal canal. The electromagnetic waves induce the development of nerve connections.

Aquarius is the fixed sign, and uses electromagnetic waves in the production of new receptors. As the sign of air follows the sign of water, so the neuro-receptors might be the imprints of antigens carried by the lymphocytes.

As the signs of the zodiac follow one another as earth, fire, water, air, the antigens connected with the lymphocytes may well carry the visual image and the emotional sensory particles as representations of the individual senses and emotions. The antigens connected with the lymphocytes may be carrying the individual representations of favorite senses and emotions in order to broaden and specify the individual personal genotype.

Gemini, the mutable sign, disseminates electromagnetic waves in the form of cations H+, which accumulate in the brain. The accumulation of cations induces the accumulation of thoughts and thought processes.

3. Light is the next element in the circle of elements in the zodiac. Light can exist in two forms: photons and waves. Light is the carrier and controls the Earth system.

Capricorn is the cardinal sign and accumulates the light.

Under the influence of light that activates vitamin D3, osteoclasts build canals in the bones that store light.

The bone marrow serves as the brain of the human being, the lymphocytes serve as the brain of God.

There is constant interaction between the erythropoietic cells, the macrophages, and lymphocytes.

The lymphocytes transfer the information and work on human receptors through the molecules of human emotions and senses.

The light held in the bones may stimulate the birth and activation of different kinds of hematopoietic cells, which are connected with different types of emotions and memories.

The erythrocytes, which come into the blood-carrying hemoglobin and its oxygen, serve the fire system as carriers of emotional, visual, and memory particles.

In this way, light serves to promote bone structure and growth.

Photons of light diminish the influence of electromagnetic waves.

Taurus is the fixed sign that uses light in the form of the photoreceptors—rods and cones. The photoreceptor cells use light in the form of waves, transforming it into light photons that reflect the image on the visual retina. The light stimulates the production of photoreceptors and increases visual images and the production of memory.

Virgo is the mutable sign that takes part in the dissimilation of light waves. The Purkinje cells of the cerebellum transform the light waves into the visual images of consciousness.

4. Oxygen serves the fire system in only one simple form, not in waves, and it carries the individual consciousness only.

Oxygen serves in the form of hemoglobin in the systems of the blood erythrocytes, which are part of the fire emotional system. The hemoglobin may be help with the synthesis and transport of emotional substances.

The fire emotional system starts in the cardinal sign Aries, in the limbic system. Aries transports oxygen through n the nose and the nasal cavities.

The molecules of olfactory and other sensory organs may be part (or precursors) of the fire emotional system.

As the carriers of the emotional substances may, the cells of the blood oxygen play a role in the formation of the emotional substances. The erythrocytes, macrophages, and the other hematopoietic cells may play the role in this process.

In all three systems of the water element, the macrophages interact with the lymphocytes, providing them with the molecules of emotional memory, which may further interact and connect with the lymphocytes and the other cells of the immune system.

Leo is a fixed sign that uses oxygen from blood vessels for the synthesis of its receptors. The Purkinje cells of the cortex may use those receptors for their function.

Sagittarius uses oxygen for dissimilation and metabolism. During these processes, it produces and stores hematopoietic cells (in the form of memories) in organs such as the spleen. These memories are stored after the interaction of blood cells with lymphocytes. The emotions and memories may be also stored in accumulated blood cells in the blood lagoons.

Sexual energy has a neurotransmitter as nitrogen oxide, which also uses oxygen as a carrier. The sexual emotions that occur after the interaction of the macrophages with the lymphocytes in the reproductive organs may be powerful contributors to the buildup of the personal genome and the selection of genes for reproduction.

Taken together, the four elements form the cycle of natural elements in the body. The elements move clockwise, stimulating each other.

THE ELEMENTS AND THEIR WORK SEQUENCE

THE EARTH ELEMENTS PARTICIPATION IN THE ELEMENTARY SEQUENCE

The work of the elements begins with the earth element, which transforms light into images. This means that the cells of all three earth systems possess photoreceptors, which transform light waves into photons.

Capricorn represents the bones and bone marrow.

The bones are the foundation of the body, and their marrow holds the progenitors of blood cells.

The bone marrow is the brain of the earth element. It produces multiple types of hematopoietic cells, which may be responsible for carrying and transporting different types of emotions and memories. The receptors for the emotional molecules may become connected with hemoglobin. Erythrocytes carry hemoglobin to meet the body's metabolic needs.

Bone density is regulated by cells called osteoclasts, which are activated by vitamin D3 and are sensitive to light. Osteoclasts are building canals in the bones, which fill with light. It could be that light waves are transformed into photons and images connected to cells found in the bone marrow.

There are reports that the bone marrow cells have the antigens similar to retinal cells, and may also possess the same or similar photoreceptors. Such cells may be able to obtain photons and accumulate sensory images that they hold and transport.

In this way, the cardinal sign Capricorn regulates construction of the bones, which serve as the primary storage area for light in the body.

The fixed sign of earth element Taurus regulates the sensory organs, which supply the person with visual images. The regulation of light in Taurus is connected with the eyes' photoreceptors.

Light stimulates these photoreceptors, which transform them into photons and images on the retina. The stimulation caused by this light activates the photoreceptors, causing them to multiply.

The mutable sign of the earth element Virgo regulates the cerebellum.

Virgo is similar to Taurus and represents the specialized nerve cell receptors—the cells of Purkinje—of the central nervous system.

There is some evidence that the Purkinje cells possess photoreceptor genes in their mitochondrial inclusions. The Purkinje cells may thus have ability to work as light-sensitive cells, transferring light into beams of photons projected into human consciousness and its visual representation.

Light is the major controlling mechanism in the earth systems.

The communicative network of the earth system moves clockwise. Light stimulation causes an increase in the number of photoreceptors, and also in the production of visual images in the sign of Taurus.

The increase in light waves stimulates the creation of bone canals by the osteoclasts, and the storage of light in those canals. The storage of images may also be possibly accomplished in the cells of the bone marrow.

The increase in light and imagery calls for their acceptance by consciousness in the sign of Virgo, which may also accumulate particular images used by the consciousness.

THE FIRE ELEMENTS PARTICIPATION IN THE ELEMENTAL SEQUENCE

The fire element takes over from the earth element, creating emotions based on information from the sensory organs.

Aries accumulates emotional experiences in the limbic system that starts near the nose and nasal cavities. The oxygen in the nasal cavities participates in metabolic processes, and is the element controlling the fire system and serving it through emotional and metabolic work.

The oxygen connects with the hemoglobin of the erythrocytes, which may carry and store emotional information through the NO (nitric oxide) molecule. It is known that the sexual organs use nitric oxide as a neuro-mediator. Nitric oxide is carried by hemoglobin, and may be fulfilling its work as a mediator for the emotional substrates (and their effects) in all fire system.

As emotions are based on information from the sense organs--molecules of smell, visual images, taste, sounds and touch--those sensory molecules may be connected with the erythrocytes as well.

As the cardinal sign Aries stores oxygen and the erythrocytes with its complexes, the fixed sign Leo uses those complexes for its work.

Sagittarius, as a mutable sign, eliminates oxygen by burning it in metabolic processes. At the same time, it acquires and accumulates emotional complexes that affect the muscles.

The fire element is regulated in the body by the humoral system, and possibly the erythropoetins and other substances that control the transportation of emotions by erythrocytes.

It is unknown, biologically, how the emotions arise. Their formation may be based on molecules produced by the sensory organs. The forming emotional molecules may undergo changes while using oxygen in aerobic metabolism, or may have interactions and connections with hemoglobin molecules.

Erythrocytes are probably the major cells responsible for holding and distributing emotional memories. Several studies have demonstrated enhanced memory in the presence of erythropoietin.

The way, in which emotional complexes influence the human body, and the mechanism of emotion-storage itself, remains unknown. It's possible these are accomplished via nitric oxide molecules working in conjunction with hemoglobin.

The communication network between the three fire signs moves clockwise.

The use of creative emotional receptors connected with oxygen (currently of an unknown nature) starts with the fixed sign of Leo. The receptors' use stimulates Aries to accumulate certain molecules of the senses and emotions—which are connected with oxygen in the blood vessels of the nasal cavities and the structures of the limbic system. This allows accumulation of certain emotional memories.

The mutable sign Sagittarius expresses emotions through the actions of the muscles. Expressing the emotions via metabolic and motor activity leads to the accumulation of additional receptors for the creative activity of Leo.

The emotional substances connect further with the signs of the water systems, and interact with the lymphocytes. Connection with the lymphocytes prevents emotional and metabolic substances from being marked as foreign bodies and targeted for destruction. Their relationship with the lymphocytes causes them to be accepted by the body.

Electrical current running through the protons of crystallized water creates a wave, transforming the personal emotional memory state into a collective one.

THE WATER ELEMENTS PARTICIPATION IN THE ELEMENTARY SEQUENCE

Cancer regulates the process of neutralizing and transforming foreign material, by using lymphocytes that interact with macrophages and hematopoietic cells to form complexes with antigens.

Cancer accomplishes the first step of accumulating and storing water, which controls the water-antigen system. The water performs its controlling functions by changing its states from simple water, to crystallized water

MEDICAL ASTROLOGY: GALACTIC CODE

connected with a complex on the lymphocyte. Multiple interleukins may be organizing the complexes for different types of lymphocytes.

The lymphocytes form complexes with the new antigens (foreign to the system) to help them acquire the characteristics of the organism—facilitating acceptance by the body.

Scorpio represents the reproductive system, and regulates the new receptors and possibly a new genotype. This work is likely to be managed by lymphocyte–type cell complexes protecting new antigens. The process may continue until sexual maturity.

Pisces represents the lymphatic system. The lymphocytes are trained in the thymus to distinguish newly-formed antigens and form complexes with them. This protects the antigens from elimination.

The thymus provides structures that serve as a template for the discovery of all possible forms of foreign material presented for protection / elimination.

The communication network between the three signs assembles them into a water system. The network's regulation moves clockwise.

As a fixed sign, Scorpio produces multiple receptors with different antigen components. The cardinal sign Cancer employs different antigen components for the absorption of food. It also transports antigens to thymus. The mutable sign Pisces checks the best possible ways of increasing the body's tolerance in acceptance of the new material, and protecting the body from foreign bodies. Pisces reaches out to Scorpio to collect and introduce more antigens, to increase the person's genotype variability, reducing the amount of foreign material for elimination.

This process of increasing the body's genotype with the help of different molecules from the senses and the nutritional and emotional antigens continues until sexual maturity at thirteen years of age. The body looks for the best genome, the best absorption of foreign particles, and a diminished necessity for foreign particle elimination. At thirteen years of age, the thymus--where lymphocytes learn to distinguish foreign from self--undergoes atrophy, and these processes slow down.

The formed complexes and the H+ waves from the crystallized

water serve as matrices to help form nervous system receptors, nervous innervations, and thoughts. Each preceding structure of the lymphocytes with an antigen complex, may serve to form the following them nervous structures. As an example, the complexes of Pisces may help to form the nervous receptors of Aquarius. The structures of Scorpio provide the way for the innervations of Libra. And the unknown structures of Cancer in the brain, possibly protecting the brain from the new food antigens, may help establishing the electric waves of the thoughts.

THE AIR ELEMENTS PARTICIPATION IN THE ELEMENTARY SEQUENCE

Cardinal sign Libra stores and uses the air element – electromagnetic current. Libra stores the anions that are the source of this electromagnetic energy in the cerebrospinal fluid. This source serves for the electromagnetic impulses, which exist only in a wave state, and carry a collective consciousness. In this way, all air signs carry collective consciousness.

The electromagnetic impulse is used in the nerve receptors of the Aquarius. The dissemination of energy takes place in the frontal lobes of Gemini, through the work of thoughts.

Aquarius represents the vast body of the vessels with neuronal endings, which belong to the neuro-vegetative system.

Gemini builds a special system of associated thoughts, and controls them for the use of electrical activity and its gradual dissemination.

The network of the systems of the air elements consists of three systems that activate each other in a clockwise direction along the Zodiac. The activation of the nervous system leads to the accumulations of the nervous receptors that are controlling the vital processes in the human body (Aquarius). The activation of the receptors requires the cardinal sign Libra to reach for and distribute more anions as electromagnetic currents in the body. That stimulates the mutable sign Gemini to increase the thought activity that accumulates more informational currents in the brain. More thought currents in the brain require the receptors to work more.

The thoughts help the formations of the senses through the transformation of the light into the photon energy and formation of the images with the help of the photoreceptor cells.

The images from the retina may increase the induction of images connected with the bone marrow as well as with the cells of Purkinje in the cerebellum, which may produce the images of the person's consciousness.

The sensory input starts a new cycle of stimulating the formation of the emotions.

IN CONCLUSION

This way the work of the elements most probably has the first impulse and starts with earth element and the organs of senses, forming the visual images. It is followed by the fire systems and forming of the personal antigens of emotions. This is followed by the water systems, and formation of the new antigens not foreign to the body. This process is concluded by the formation of the nervous system that is repeating the matrix of the formed complexes of the lymphocytes and the proton H+ waves.

The sequence of the four elements repeats three times.

First comes the formation of the emotions, which concludes with the organization of the nervous receptors throughout the body in the sign of Aquarius.

Second is the process of the formation of the relationships, which is concluded with the formation of the nervous organization of the body in the sign of Libra.

Third is the process of the formation of the central nervous system, which is concluded by the special organization of the nervous currents, creating conversational thoughts in the sign of Gemini.

Each element relies on the prior element's work for its foundation, and serves as a foundation for the next one.

As in the network connection of the systems within one element, the sequence of the work of the elements moves clockwise.

12

MEDICAL APPLICATIONS OF THE HUMAN ZODIAC

We have now described the signs that code and regulate different areas of the brain, the human senses, and the main parts of the nervous system—all of which play a part in regulating portions of the body and the activities of organs. We have also described the signs that code the main body systems and the general systems that are necessary for walking and leading an active life.

Seven signs are positioned on top of the chart and form a half-ecliptic around the person's head like a beautiful wreath. The other five signs form constellations around the torso and the legs.

While the signs code the energy of the person's systems, planets in the native chart construct a personal blueprint of energies and carry the expression of these energies in the psychological and biological phenotype of the individual.

It seems that information about the bodily structures downloads during the embryonic stage of human development and may not depend on the moment of the person's birth. The download for the psychological and biological energies depends on the location of the planets at the moment of birth.

As the planets in the chart bring the energies of the signs they rule, the nature, position, and interaction of planets bring up the planet's energies for medical issues as well.

It seems the luminaries—the sun, the moon, and Jupiter— increase the energy of the sign they are in and decrease the energy of the sign on the opposing sign of the axis.

A similar effect might be generated by other planets; influencing the signs they're in as well as the opposing signs.

The correlation of planets' distribution in the human chart with the human-system energy state might form a predisposition toward or vulnerability to disease. This predisposition needs to be scientifically examined before any definite correlation can be announced. The astrodata of different correlations with human diseases may be placed in a computer for statistical analysis, after which it may be used for medical and prognostic goals. The example here is to demonstrate how the logic of the system works. The correlations have to be worked out in a proper way before being employed in readings.

In this chapter, we examine a correlation between the distribution of the planets in the astrological chart and kidney disease, using as our example Alexander III, czar of Russia. Alexander was a healthy and vigorous man known for his strong energy and body type. While young, the czar developed an infection of the urinary tract and died from nephritis.

The sign that rules the kidneys and the nervous regulation of the kidneys is Libra, and the planet transferring this energy is Mars.

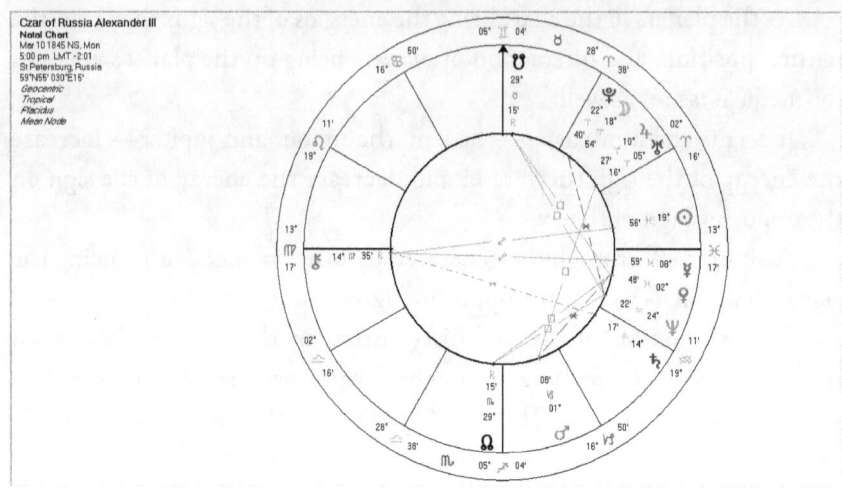

We can see from the chart that the sign of Aries is overloaded with planets, including the luminary moon. The planets contribute to the energy of Aries, leaving Libra (the opposite sign) with less energy and stability.

Mars obtained its position in Capricorn, contributing to the known strength, physical activity, and influence that the czar had. Being in the Earth energy of Capricorn, Mars squares the energy of Libra.

At the same time, Mars squares the planets in Aries and so has a major influence on the czar's energy, shown by Jupiter and Uranus. In this way, Mars is involved in the T-square that inhibits its energy as well that of Libra (which Mars rules). The sun's location in the seventh house of Libra may activate the conflict.

The location of Mercury in the sixth house may have influenced the nervous system in general. The location of Venus in the same house may have conferred a predisposition for diabetes, as Venus influences the taste buds.

The location of Neptune in the sixth house (opposite its own twelfth house) may have influenced the decreased function of the lymphatic system (ruled by Neptune).

Overall, the chart seems consistent with the czar's vulnerability to kidney disease.

THE INFLUENCE OF THE HUMAN ZODIAC ON THE DEVELOPMENT OF THE GLANDS IN THE BODY

Each sign and its ruling planet are responsible for creating a gland in the human body. This work enables the created gland to satisfy the function of the sign that creates it and serves the function of the next sign and the functions within the whole body. When I refer to the next sign, I mean the next sign of the zodiac, following clockwise, since we presume that the development of consciousness follows the clockwise direction over the zodiac signs.

We start with the sign of Cancer, which is responsible for the first organisms, which developed in the water. The other signs that belong to the water element and were involved in the creation of the first living organisms are Pisces and Scorpio. As all the life in water was influenced by the outer planets of the solar system, those signs are ruled by Neptune and Pluto, respectively. Thus the first organisms were mostly engaged in secretional functions regulated by the sign of Cancer. Besides secretion, they possessed the protective function (Pisces) and reproductive function (Scorpio).

THE DESCRIPTION OF THE FUNCTIONS OF THE SIGNS BELONGING TO THE WATER ELEMENT

Cancer – involved in building the stomach, the major metabolic organ of the first multi-cellular beings. It also provides the sensory cells—neurons of the pineal gland responsible for providing the signals for day and night.

Pisces – (ruler – Neptune) involves multiple cells in the front of the body in the protection and secreting chemicals in order to recognize each other for assistance and interaction (secreting interleukins in the human body).

Scorpio – (ruler – Pluto) involved in the reproduction cycle, which is mediated by the sexual hormones for purposes of the maturation of the egg, its ovulation, and division.

Aquarius – (ruler – Uranus) builds the organs connected to the neuro-vegetative system, vessels carrying neuronal receptors of the vegetative neurons system. Aquarius helps in creating the multi-cellular organisms and low vertebrates, most of which are connected with life in water. Uranus is involved in the building of the parathyroid gland, which is necessary for the work of the next constellation, Capricorn.

Capricorn – (ruler – Saturn) uses the parathyroid hormone in building its major structures, bones, as well as ligaments and joints. Those structures bring to life higher vertebrates, some of which attain large dimensions (dinosaurs). Capricorn is involved in secreting synovial fluid, which is used in the creation of the knee, which is also used by the constellation of Sagittarius for building its major joint construction, the hip joint.

Sagittarius – (ruler – Jupiter), gives rise to the influence of all planets of the solar system. It's located at the center of the Milky Way, to which our solar system belongs. At this stage, the living being acquires the more sophisticated systems of metabolism, hematopoiesis, neurotransmitter influences and metabolic and secretionary organs like the liver. The endothelial cells of the vessels cooperate with the protective cells (lymphocytes) to build the hematopoietic organs, which produce the blood cells (spleen, lymphatic, nodes).

It is also possible that Sagittarius is connected with the development of red blood cells' oxygenation and the process of metabolism.

Sagittarius influences the development of the more sophisticated organization of the brain/neocortex.

With the action of the constellation of Capricorn, the higher vertebrates appear. They have large bones, large joints, and accommodations for movement on the earth. The hip joint maintains the leg in its vertical position, and the belly rises above the level of the earth, being supported by the legs. (This is a hallmark of evolution.)

The vertebrae carry the spinal brain, which is supplied with the energy of kundalini rising from the nervous nuclei, which are located in the spinal columns of the spinal brain. These give energy for movements.

Sagittarius gives rise to the pituitary gland, with its production of follicular stimulating hormone, oxytocin, prolactin, and adrenocorticotropin-stimulating hormone (ACTH). The next constellation, Scorpio, uses these hormones for reproduction, delivery, and lactation.

Scorpio – (ruler – Pluto) gives life to living creatures, which use their hormones for maturation, proliferation, and intrauterine reproduction.

The pituitary gland also produces thyroid-stimulating hormone (TSH), which the constellation of Gemini (ruler – Mercury) uses in the building of thyroid gland, which produces the thyroxin hormone, which influences the metabolism of the body.

Control over the hormonal regulation of the body belongs at this stage to the hypothalamus, which secretes the releasing factors that help in the secretion of the pituitary gland.

Cancer participates in the process of intrauterine pregnancy and lactation by having regulation over the hormone prolactin and secretion of milk after the end of pregnancy. Cancer may as well have control over the basal ganglia that secrete dopamine, serotonin, acetylcholine, and norepinephrine. These substances are important for the regulation of the body as well as different parts of the brain.

ACTH is produced by the pituitary gland is used by the constellation of Libra, which codes the kidneys with the adrenals.

Libra – (ruler – Mars) rules twelve nerves that come out from the base of the brain. The nuclei of the nerves are contained mostly in the medulla oblongata, as well as in the pons.

Libra is part of the neuro-vegetative system of the air element. The neural cells participate with their axons to form the neuro connections that activate the relationships between the body parts and organs. Libra participates in the innervating of the adrenals, which create the fight-or-flight response.

Libra is connected with the development of mammals. The intrauterine development that is characteristic of mammals also appears with the development of the reproductive organs innervated by the sign of Libra.

Adrenals participate in the construction of the gender-sensitive cartilage of the throat (Adam's apple), coded by the constellation of Taurus (ruler – Venus).

Both Mars and Venus participate in creating the human voice. The ability to talk and sing is unique to human beings. Mars participates in building the cartilage rich in chondroitin-sulfate.

Taurus – (ruler – Venus) participates in building the thymus gland that protects the throat immunologically and becomes the foundation for the maturation of the lymphocytes, regulated by the constellation Pisces (ruler – Neptune.)

The lymphocytes are the cells that protect the organism and in the process of their development undergo both positive and negative interactions with the cells of the body as well as with foreign cells. Both are necessary outcomes of being so close to the foreign bodies that the difference would be only in one antigen construction. That would allow them to completely merge with the foreign cells and eliminate them from the body, in this way serving the higher form of love.

The next constellation in the zodiac going clockwise is Pisces, which uses the thymus. The role of Pisces is protection of the body. The thymus is a main lymphatic organ that teaches and produces the mature lymphocytes.

The thymus, as an organ, is active only until the sexual maturation of the person, or about thirteen years of age, marking the end of the childhood individual development. It is interesting that Neptune, the ruler of Pisces, remains in a single sign for about thirteen years. It marks the age of the person who is at the end of the phase of development of the individuality that contributes to the generational and collective unconscious. Pisces therefore is the bearer of the collective and generational emotional unconscious.

Upon sexual maturation, the thymus stops its work and disappears, and the preferential work shifts to the constellation of Scorpio and the action of the sexual hormones.

Pisces – (ruler – Neptune) creates the interleukins and the many factors used by the next constellation, Aquarius (ruler – Uranus), in the creation of the hematopoietic organs (organs that create the blood cells—the spleen, the liver, lymphatic, nodes). Thus the constellations along the human ecliptic serve each other to fulfill their functions.

Aries – (ruler – the moon) is the carrier of the human emotional consciousness in the emotional brain, which is represented by the limbic system and basal ganglia. Aries may have responsibility over the basal ganglia pertaining to early childhood development. This development strongly depends on all learning and emotional experiences that the child gets from his close family. On the level of the central nervous system, the constellation of Aries is connected with the constellations of Libra and Taurus.

The sign of Libra, which is responsible for forming relationships between people and their quality, lies opposite the sign of Aries and establishes the level of stability and balance that forms in the basal nuclei of the child.

The formation and interactions of some basal ganglia depend on the interaction of the opposite sexual hormones (regulated by Mars and the moon), which determines the size and connections of the basal ganglia in the thalamus.

The hormonal regulation and nervous connections of the basal ganglia in childhood become a factor that determines gender identity and sexual orientation. In homosexual orientation, the areas around the basal ganglia have been reported to have more neuronal connections and white matter around them that has some different paths that don't usually form in heterosexual individuals.

The role of Aries in human development is to bring human consciousness to the highest level by increasing the maturation of the nuclei. Aries gives signals to the whole body through stimuli from the basal ganglia, hippocampus, amygdala, mammillary body and all the other structures belonging to the limbic system. The sense of touch is delivered in human emotional consciousness by way of interaction with the sign of Libra. Taurus provides the human senses located on the face and the ability to talk. Those abilities of the senses create a foundation for the human emotional brain.

Gemini – (ruler – Mercury) presents an unprecedented example of a system in the body that acquires total consciousness over itself. Gemini gives the human being the possibility of conscious determination of each step and creativity in one's goals and destination.

Leo, Cancer, and Gemini carry the responsibility of creating the part of the brain in contact with the celestial spheres and the upper body of the human, the chest and its organs.

Gemini carries a large responsibility for the upper part of the body, creating the lungs, ribs, shoulders, arms, and hands. The arms and hands, especially the fingers, have their own representation in the area of the brain that is ruled by Gemini, which allows them to create any movement immediately after the directive thought appears. For example, when a person does something with his hands, such as writing or working with clay, a thought is immediately reflected in the action of his hands right away. The face has the same representation. Gemini represents the space in the brain that controls the parts of the face, allowing facial movements to occur in quick response to actions and thoughts.

At the opposite side of the axis, at the constellation of Sagittarius (the sign of Sagittarius lies across from Gemini, on the same axis), the movement is produced with the muscles of the legs (kundalini) and requires time for the nervous signal to travel from one representational area of the brain to the other. The delay happens because the organs regulated by those constellations have separate representative areas in the brain. The signal from the nerve ending travels to one area of the CNS and then has to connect with the nervous system that represents the part of the body that has the muscle.

The constellation Gemini codes the thyroid gland, which is also regulated by the sign of Sagittarius (which rules the pituitary gland). This gland provides the energy and capacity for the metabolism of the whole body. It helps with thermo-regulation and affects the respiratory center and cardiovascular activity.

Cancer – (The ruler not assigned in this system. The rulership of the moon moves to Aries. Sirius may be the ruler) has the influence on the basal ganglia regulating the vital body functions. The stomach of a person may represent a system that responds to thoughts unconsciously. The hormones secreted by the stomach may mediate emotional eating, which is a form of responding to emotions. Digestive hormones serve to produce a supply of energy for the whole body.

Cancer regulates part of the brain cortex and the super-conscious part of the mind that may be represented by the pineal gland.

Leo – (ruler – the sun) reflects the unconscious part of the cosmic consciousness. It codes the parietal, temporal, and occipital parts of the brain as well as the heart and the spinal brain.

The right auricular myocardiocytes of the heart secrete a peptide hormone—cardiodilatin—that regulates the vascular smooth muscles. The secretion of this hormone may influence the regulatory functions of the constellation of Virgo by influencing the smooth-muscle endothelial cells of the cerebellum.

Virgo – (ruler – Chiron) the only sign that belongs to the element of earth that is ruled by an outer planet, Chiron, which lies beyond Saturn. This is to say that in the highest of its achievements, the human kingdom is of God and does not belong to the solar system particularly. Both planets that rule the element of earth—Saturn and Chiron—lie outside Jupiter that represents the beginning of the fire system, that might be regulated by our solar system.

Saturn, the planet that started the existence of people, is located between Jupiter and Neptune, which represent the generational memory and collective unconscious. That could mean that Saturn is giving birth to a person who will give birth to the collective unconscious.

Chiron codes the oldest part of the brain, the cerebellum, and rules the constellation of Virgo. Chiron rules the secretion of the pancreatic gland and the metabolism through the path of glycolysis, which provides the substrate for the function of the brain. Virgo has been shown to participate not only in the motor-stability function of the brain but in the analytical functioning and some important functions of the body, giving stability to the whole organism.

Virgo is thus at the entrance of the brain and leads to the signs that code the top of the brain. It also regulates the metabolism, providing the nutritional substrate for the brain.

Meanwhile, Gemini is located at the exit of the brain, leading to the signs that code the emotional brain and the body and providing the path for the metabolism and the nutritional support for the body.

TO SUMMARIZE:

Cancer provides the body with the pineal gland and the stomach as secretory organs.

Gemini builds the thyroid gland (with the help of Sagittarius), which provides necessary energy for the whole body.

Aries provides the mediators for emotional support, influencing the human emotional consciousness and the amygdala, with its memory.

Taurus provides secretions for the sensory organs of the face and for the throat and builds the thymus for the immune protection of the organism.

Pisces uses the thymus for immune protection of the whole body. It regulates the secretion of the multiple interleukins for construction and protection of organs such as the spleen and lymph nodes that are responsible for blood differentiation and interaction with the cells of the immune system.

Aquarius builds the neuro-receptors of the blood vessels. It regulates the parathyroid gland, which is used for creating the skeleton and the bones of the higher vertebrates.

Capricorn uses the parathyroid gland for the creation of the vertebral column and the bones. It provides the secretion of the synovial fluid in the knee joint.

Sagittarius uses synovial-fluid secretion for lubrication of its major joint, the hip. The work of Sagittarius is connected with the large muscles that require intense metabolism for energy production. Sagittarius codes the pituitary gland, which regulates the secretional production of the thyroid glands, the steroid sexual hormones, and the corticosteroids, secreted by the adrenal cortex.

Scorpio responds to the pituitary-gland hormones by secreting the sexual hormones and using it in its hormonal cycle. The sexual hormones produced by Scorpio are used in the constructive functions by many organs and of the body.

Libra balances the level of secretional production of the adrenals, kidneys and other internal organs regulated by the vegetative nervous system. The name *Libra* is proposed because of the feedback that the sign offers in organizing and providing balance to the work of the internal organs. Libra responds to the pituitary-gland hormones with the secretion of the male hormone testosterone from adrenals and male sexual organs, as well as corticosteroids from the adrenals. Testosterone is used for interaction with the gender-specific thyroid cartilage and for construction purposes by many organs.

Virgo is influenced by all the mediators influencing the brain cortex, which are produced by the basal ganglia of the brain, and by the cardiodilatin, which is regulated by the sign of Leo. Virgo is thus more connected with the signs that code the top of the brain. Virgo builds the pancreatic gland and provides itself and the three leading constellations of the cortex—Leo, Cancer, and Gemini—with glucose and vital energy through glycolysis, which occurs mostly in the liver. The enzyme needed for the distribution of the glucose is insulin, which is produced by the pancreas. At the same time, Virgo maintains the motoric functions of the body and its stability.

Virgo is located near Leo, which is ruled by the sun, and of all the constellations, it alone is represented by a human face—that of Persephone, who in myth lives half of life on Earth and half of life with Pluto in the celestial spheres. As ruled by Chiron, the human has the taste of another kingdom and belongs to a divine realm. The cortex of the cerebellum, which has immense dimensions in its involuted form, can bring multiple functions and transformation to the whole human being.

The consciousness of the human in the zodiac is next to the consciousness of the sun.

Leo uses the metabolic process provided by Virgo. It supplies the body with multiple neurotransmitters and regulates the cardiovascular system. Leo uses oxygen for the accumulation of its receptors and takes part in the work of the emotional fire system.

Using the path of the interactions of the zodiac signs, we now come back to the constellation of Cancer, which partakes in the consciousness of the whole Zodiac sign and transforms it into the energy and pure consciousness of the Gemini, which leads to the new ecliptic of the zodiac path, where all the constellations interact in producing the human consciousness, human emotions, the human life, and human interactions. The zodiac is the circle of life for the human being.

Afterword

My connection with astrological medicine began with the study of my own astrological chart. In that chart, the ascendant is at the end of Sagittarius – 29° forming one angle of two trines, one in Fire and one in Earth. Those are the points of self-construction and self-expression in the world, where the person walks his earthly path.

I had my first insight that those two signs might be connected.

The meeting of the Fire and Earth triangles on the Earth plane happens in the connection between Capricorn and Sagittarius. The logical connection of these signs made me think about dividing the signs on the Ecliptic into functional trios.

The trios of constellations regulate function on the earthly plane. On the celestial plane, two other trios connect the earthly and celestial realms: the emotional trio (on the left side), and the relationship trio (on the right side).

Additionally, the trios represent themselves as the parts of the human body. The main horizontal axis—connected by the signs Libra and Aries—represents the precession of the equinoxes. These separate the constellations (which code the parts of the brain that include the constellations Libra and Aries) from the rest of the body.

The four elements began to reveal themselves to me. Each element included three astrological signs that represented one full bodily system (itself composed of three parts, each coded by one constellation).

Each system typically consists of the leading, transitional and final (or analytical) part. These, in turn, correspond with the cardinal, fixed and mutable signs of the zodiac. The mutable sign represents the end part of the biological system, which analyzes the entire system to determine whether any goal tasked to the system has been achieved.

The signs, also, started to reveal themselves.

Medical astrology assumes rulership of the planets according to the function fulfilled by the sign and the planet, and is supported by the length of the planet's return, which provides energy for the function.

Medical astrology provides information about the creation and coding of the human body. It provides insights into the evolution of life on earth. This data provides a whole new horizon for further research.

Knowledge of biological design is of medical significance, as planetary interactions reveal how the energies coding the medical systems interact. For example, some months may be better than others for medical and surgical interventions and systemic wellbeing.

Familiarity with the psychological energies provided by the planetary zodiac surrounding each individual (the personal zodiac) is a major tool in psychology, and helps determine the subject's major potentials and conflicts.

Information contained in the astrological chart may be programmed into a computer, and used for diagnosis, prophylaxis, and treatment in the medical and psychological sciences.

Acknowledgements

The book is dedicated to my parents, who have always trusted in the wellness and richness of the world.

This book wouldn't be possible without the help and inspiration from Derek O'Neill.

I am sending gratitude to my family – my husband Vladimir and daughter Julia who were always supportive of my work.

www.ingramcontent.com/pod-product-compliance
Lightning Source LLC
Chambersburg PA
CBHW051545170526
45165CB00002B/891